Automation and Basic Techniques in M
Microbiology

Santi M. Mandal • Debarati Paul

Automation and Basic Techniques in Medical Microbiology

Springer

Santi M. Mandal
Central Research Facility
Indian Institute of Technology Kharagpur
Kharagpur, West Bengal, India

Debarati Paul
Amity Institute of Biotechnology
Amity University Uttar Pradesh (AUUP)
Noida, Uttar Pradesh, India

Both authors contributed equally

ISBN 978-1-0716-2371-8 ISBN 978-1-0716-2372-5 (eBook)
https://doi.org/10.1007/978-1-0716-2372-5

This Springer imprint is published by the registered company Springer Science+Business Media, LLC part of Springer Nature.
The registered company address is: 1 New York Plaza, New York, NY 10004, U.S.A.

Dedication

This book is dedicated to every tree whose life I have shared. You have increased my appreciation for the terra firma under my feet and the heavens you are forever reaching towards. You have made my life – personally and professionally – a daily wonderment of observation and exploration and inspired me to share that pleasure and privilege with your human associates.

And, just as trees do not exist in a void, neither do people. It was my good fortune to enter into the world of Urban and Community Forestry at a time when a whole host of caring, knowledgeable urban forest professionals were emerging – from the USDA Forest Service Durham Field Office: Sharon Ossenbruggen (RIP), Bernie Raimo, Bob Neville, and Dr. Alex "Touch Trees" Shigo (RIP) – (Sharon and Alex, Sharon's next-door neighbor, were the motivating force behind my becoming a Consulting Arborist) as well as Peter Frank from the NYS Department of Environmental Conservation (NYS DEC). They were more than willing to embrace an enthusiastic and dedicated "newbie." No creative idea was too farfetched to explore and invest in. They independently and collectively offered support and guidance and encouraged my professional development into the Urban Forester and Consulting Arborist I became. Without their generosity, which continues to this day through the USDA Forest Service i-Tree initiative under the leadership of Dr. David Nowak, this would not have been possible, and I remain in their debt and dedicate this book to them as well.

Epigraph

"A thought comes to mind of summertimes ago, of trees in Central Park,
children playing Tarzan, making believe branches were tree houses and that we were free...

Trees, trees, trees without end, but if that is to be true, we must be friends of trees so they will grow to be mighty from roots that are strong. It's the same for all the children in a world without end. Amen."

Piri Thomas, 1993
in honor of NY State's
Urban & Community Forestry Program

Preface

I began my love affair with trees some 50 years ago. Growing up in a family of limited financial means in the east Bronx, a borough in the City of NY, had its constraints. Life in a tenement, while culturally rich, did not include visits to any natural areas beyond what was immediately available, which was not very much. The east Bronx neighborhoods of those years did not reflect the urban greening largesse that was made available to wealthier neighborhoods in Manhattan and Brooklyn, thus our streets were predominantly brick tenements and grey pavements. There was quite a bit of experiential nature at home, with a glorious English cocker spaniel, my father's tropical fish breeding project, and my injured critter hospital. Forays into the *wild* consisted of my father taking me to the New York Botanical Garden and the Bronx Zoo. These cultivated places were the extent of my *natural* nature dabblings during my formative and early teen years. It wasn't until my husband-to-be suggested a hiking/camping excursion to the Catskills Mountains, a protected *Forever Wild* section of New York State, that I experienced trees in a forest wilderness for the first time, an experience that was to alter the direction and focus of my life.

The smells, tastes, and sounds filled all my senses. I lay down on the ground, looking up at the cathedral framing the sky, felt the pulsing of unfettered life beneath my back. I was embraced by the forest around me, and when I returned the embrace by putting my arms around a red oak, the forest's energy that was contained in that tree along with all its magic and mystery filled my being and I was metamorphosed.

When you enter a grove
peopled with ancient trees,
higher than the ordinary,
and shutting out the sky with their thickly inter-twined branches,
do not the stately shadows of the wood,
the stillness of the place
then strike you with the presence of a deity?
– Seneca

That experience has led me along many tree-oriented by-ways. Over the passing decades, it became more and more clear that if both trees and people were to thrive, people needed to embrace the entirety of this organism – what it is, what it needs, and how can we best provide that. This is especially true in this time of great environmental upheaval in which we, as a species, have exceeded the capacity of our planet to sustain our personal and collective well-being.

As climate change accelerates and the prognosis becomes more and more grim, one wonders if the outcome is written in stone? How is everything we did yesterday not only impacting the world of today but what it means for the world of tomorrow? Can we no longer close the environmental Pandora's box that we have opened?

While the problems seem daunting, there are actions that can be taken by each of us in our daily lives that can and will make a difference. There are more and more of us asking that most simple and poignant of questions *What can I do…?* It is the sum of all the small actions and initiatives that can change the story's outcome.

The raison d'être for this book, this homage to the indomitable spirit of a Tree, is to capture the world of Tree in all its aspects: its biology – what makes a tree a tree; its place in myth, art, poetry and prose; its profound connection to us; to make all of that knowledge actionable and to offer a variety of *What can I do?* answers; to show how this sentient and glorious entity has inspired informed stewardship around the globe and how that stewardship can mitigate an aspect of the climate change impacts we have set in motion and, at the same time, reconnect us to each other and to the earth that sustains us.

Luzern, Switzerland Naomi Zürcher

Acknowledgments

Life, its products and by-products, never exist in a void. It is all an actionable collaboration, a composite of an extensive array of connections. Just as the development and realization of my practice under the alter ego, Arbor Aegis, was facilitated by many collegial partners and my wonderful life partner, Al Zürcher, so too was this book project. This acknowledgement and most heartfelt *thank you* is a small abbreviation of the extent of my gratitude for making the journey possible.

Researching and crafting *Connecting Trees with People* was a fascinating exploration and an endless metamorphosis of thoughts and ideas. Such an endeavor always takes a village, and the book's village is as follows:

- All of those who took time from their busy schedules to share their creative and motivating participatory programs – Beirut LB: Adib Dada; Bhutan: Kaspar Schmidt; British Columbia CA: Cecil C. Konijnendijk; Cairo EG: Abdallah Tawfic; Celje SI: Robert Hostnik; Chicago IL USA: Taykhoom Biviji; Concepción GT: Jorge Rodriquez; Edmonton CA: Lisa Fialka; Greenpoint NY USA: Sarah Balistreri, Acacia Thompson; Lagos NG: Doyinsola Ogunye; New Delhi IN: Akshat Tyagi; Oakville CA: Anelia Tichkova; Padua IT: Thomas Campagnaro, Simone Iacopino; Puerto Rico USA: Elvia Meléndez-Ackerman, Chris Nytch; St. Gallen CH: Florim Sabani; Säo Paolo BR: Maurício Lamano Ferreira; UK: Kieron Doick; Unley AU: Jenni Garden; Victoria Park AU: Heather Johnstone; Washington D.C. USA: Jessica Sanders; York Region CA: Michelle Bourdeau; Zagreb HR: Marijo Spajic.
- All the artists, both friends and acquaintances, who hold trees and their landscapes dear to their hearts and to all of those artists who freely shared their inspiring visuals – Kathrin Blümli, Barbara Davi, Margarit Eberle-Jorg, Valentin Egli, Erin Hauer, Pablo S. Herrero, Katie Holten, Nils Nova, Pejac, Judy Shigo, and Julia Sommerfeld. And, to those who generously offered cost consideration, from David de la Mano and Pablo S. Herrero for use of their extraordinary *Birds Time* as cover art and then donated an additional piece, *Sindrome del Espejo*, for the very last page in the very last chapter of the book to Kathrin Blümli, who was able to interpret my

idea of visually expressing the global extent of sacred trees and produce *A Global Sampling of Sacred Trees* – a wonderful piece of original art in Chap. 1.

- Last but certainly not least, friends/colleagues who gave their time to read the chapters, sharing their invaluable thoughts and suggestions – Jane Carstairs, Erin Hauer, Robert Hostnik, and Alonzo Zamora.

Abstract

In our ever-evolving relationship with nature, urban living has initiated what was once thought of as an oxymoron – the urban forest – a complex ecosystem consisting of trees, associated green spaces, and all their living inhabitants, including us.

The journey has taken us from once verdant forests, as described in John Fenimore Cooper's *The Pathfinder*: "It was the vastness of the view, the nearly unbroken surface of verdure, that contained the principle of grandeur…, induced a feeling allied to awe."[1] Trees growing in a forest community, as far as the eye could see, and now there's us, with our relentlessly expanding and urbanizing footprint. Joni Mitchell captured the essence of those urbanizing actions in *Big Yellow Taxi* with that famous refrain "Don't it always seem to go that you don't know what you've got till it's gone. They paved paradise and put up a parking lot."[2] Extensive research combined with public health and quality-of-life issues have reinforced the realization that we cannot exist without putting back into our denatured urban environments a part of the nature that we so definitively removed.

With increasing concerns related to climate change, the urban forest journey – putting nature back into our urban areas – has not only continued to evolve over the last five decades but has taken on much greater importance. The urban forest has become an essential ingredient in creating a healthy sustainable urban ecosystem, thanks to an extensive body of knowledge from well-documented research and practical applied experience but also because it grew, in part, out of a grassroots effort – bottom up, not top down. It is inclusive, not exclusive, and it's growth was thanks to the potential of a broad-brush approach to planning and implementing such a forest. It is quintessentially URBAN if you define urban to include people, and it is people – stakeholders – from all different perspectives that have made urban forestry live wherever its seeds have been planted.

[1] Cooper, J. F. The Pathfinder, part of Leather Stocking Tales. 1840.

[2] Mitchell, J. Big Yellow Taxi, written, composed and recorded 1970.

This book is about that aspect of the urban forest journey which has enabled the vital connection between urban trees and their human associates – an urban tree odyssey that

- Offers a basic understanding of what makes a tree a tree – biologically and structurally
- Explores the philosophical and cultural underpinnings of urban forestry
- Examines the implementation of that philosophy and the resulting "greening" practices which have taken seemingly urban-tolerant trees from forests all over the world and placed them in our urban centers, often in conditions even these resilient and tenacious organisms cannot withstand
- Investigates the global initiatives that have afforded interested and concerned citizens an entrée into the extraordinary world of urban trees
- Analyzes the capacity of these initiatives to grow the urban forest from the perspective and experience of an urban forester/consulting arborist.

Contents

List of Figures

List of Tables

Chapter 1
Introducing the Tree…

Introducing the primary character in the book – the tree: in biological and structural terms as well as its symbolism in myth and the arts.

Keywords Architecture · Compartmentalization · Ecosystem services · Growth · Mythology · Oxygen · Photosynthesis · Structure · Symbolism

Fig. 1.1 John Muir/Pacific Crest Trail, 1976. Naomi Zürcher

N. Zürcher, *Connecting Trees with People*, Future City 16,
https://doi.org/10.1007/978-3-030-94534-3_1

The tree is more than first a seed,
then a stem, then a living trunk,
and then dead timber.
The tree is a slow, enduring force straining to win the sky.
~Antoine de Saint-Exupéry, *The Wisdom of the Sands*,
translated from French by Stuart Gilbert,

Since Homo sapiens first appeared on this earth some 200,000 years ago, the road we have traversed has been very long and very diverse. The web of critical relationships we formed over these many millennia has been and continues to be filled with trees. Our mythology, our art and our poetry have celebrated the tree and its connection to us. So let us set the stage for the modern day continuity of this homage by starting at the beginning with a basic understanding of the what – what makes a tree a tree.

Oxford's online dictionary defines a tree as *A woody perennial plant* (lasting or existing for a long or apparently infinite time), *typically having a single stem or trunk* (trees can also be multi-stemmed, with multiple trunks clearly visible just above the ground). They grow to a *considerable height* (location can cause stunting, e.g. Bristlecone Pine [*Pinus longaeva*]) and bear *lateral branches at some distance from the ground* (there are many species whose lateral branches touch the ground, e.g. American Beech [*Fagus grandifolia*] and will then root, forming colonies of the tree).

It is clear from Oxford's definition and the stated inconsistencies that attempting to capture absolutes when dealing with sophisticated living organisms ignores the unique aspect of individuals within species and genera. Rather, as this homage is all about connections, let us look instead at similarities and differences to establish a working understanding, especially as it relates to how a tree connects to everything it coexists with.

Basic Tree Biology

Trees and people have life's essentials in common. We are both part of a system and we both require air, water, nutrients and energy to fuel the system in order that it may live and grow. These are the basics of our shared must-haves but it's what distinguishes us that's at the heart of the *what* that is a tree:

- humans, like the rest of the animal kingdom, do not produce their own food. They rely on the plant kingdom to directly or indirectly provide the basis for the nutrients we depend on. Trees are a completely dynamic system, relying on materials they themselves produce. As a result, trees do not grow beyond their ability to support themselves.
- humans, like all other mammalian species, are capable of regeneration. If we are wounded or if certain organs are damaged, we have the capability to repair, restore and/ or replace damaged cells or tissue at the very same place as the site of the dam-

age. Trees cannot RE-generate at least not in the same way regenerating organisms can and do. Rather, trees generate. Trees make new cells in new places. They grow in layers of cells that are actively dividing, basically adding a new tree every year. (Fig. 1.2 Dr. Alex L. Shigo. *Modern Arboriculture: A Systems Approach to the Care of Tree and their Associates*. Durham NH: Shigo and Trees, Associates, 1991, 72.)

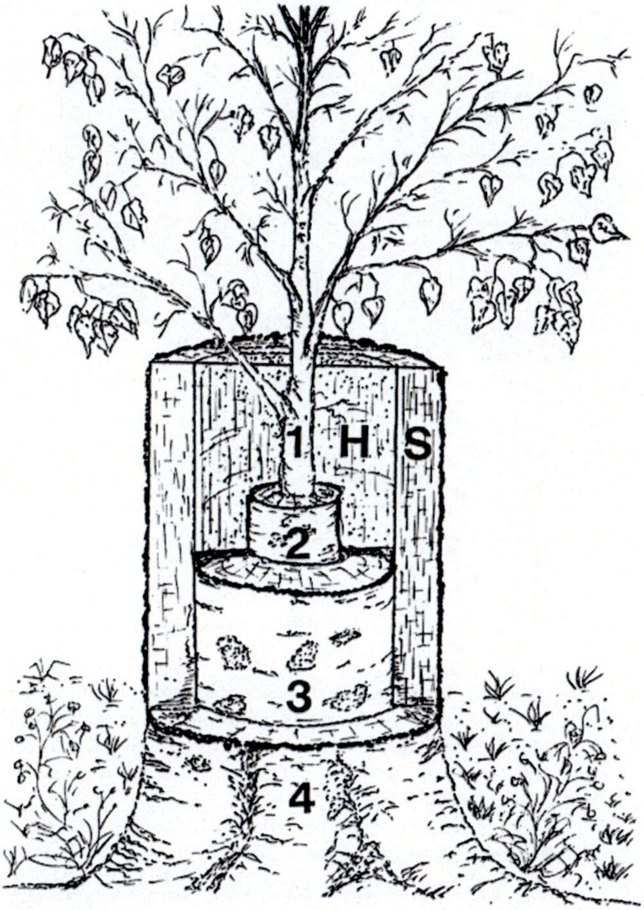

Fig. 1.2 A multiple plant
Every growth period a "new" tree grows over the older trees: 1, 2, 3 & 4. H = heartwood and S = sapwood. Shigo (1991, pp. 72)

These layers are highly compartmented, structurally bound by rigid cell walls. (Fig. 1.3 – Ibid., 71.)

Fig. 1.3 Compartmentalization model
It was time to revisit and re-examine the tree. Organisms develop in ways that suit their survival. Maybe the tree has some different ways for survival after injury and infection? Maybe the tree is not just a large core of dead wood surrounded by a single layer of living cells, the cambium? Can trees heal? Healing was the crucial question. From all of these questions, a "new" type of tree began to emerge, the compartmented tree
The tree was seen as a highly compartmented system. Each growth ring was a compartment, and each growth ring was divided into smaller compartments by radial sheets of parenchyma cells [R] The cambium was not a single layer but a cambial zone [C]. P is the Pith. Shigo (1991, pp. 71)

This growth process defines how trees deal with such things as mechanical injury or necrotic tissue. The strategy is known as compartmentalization and the process is one of containment, a process that must be maintained for the life of the tree. In other words – trees don't heal, they seal.

- humans maintain their core body temperature by sweating when it gets too hot; sweat is comprised of water and sodium and it cools us temporarily as it evaporates; this self-regulating strategy only cools the individual who is sweating.

A tree's cooling strategy serves a number of purposes. Several processes within the soil and the tree enable evapotranspiration – the transpiration of water with its nutrient accompaniment from roots to shoots and evaporation of that water, primarily from leaves, as a cost of the diffusion of carbon dioxide through open stomata for the purpose of photosynthesis. As water is released into the atmosphere, it vaporizes and it is this water vapor that quantifiably cools the surrounding air.

In addition to these functional differences, there are those differences in which each species' end products are critical to the survival of the other as in cellular energy and cellular respiration. Simply stated, while we both require the production of cellular energy and we both undergo cellular respiration, the approach is quite different.

We get our cellular energy – our system-wide fuel – from the food we eat. The glycolytic system breaks down the carbohydrates present in digested food to form usable chemical energy, primarily in the form of a sugar – glucose. Human cellular respiration converts that glucose into CO_2 and water and, through the breaking of those bonds, releases life-supporting energy in the form of ATP. The aerobic form of cellular respiration in humans requires the presence of oxygen in the process.

Trees, like all plants, produce sugars quite differently. Through the process of photosynthesis, trees use the energy their leaves and stem(s) harvest from sunlight to convert atmospheric CO_2 plus water into glucose plus oxygen as a by-product, written as $6CO_2 + 6H_2O > C_6H_{12}O_6 + 6O_2$. In order for trees to convert the glucose they produced into usable cellular energy, oxygen, a necessary element, is taken up by leaves, stems and roots.

As both oxygen and CO_2 are critical to the survival of not only trees but humans as well, the need to support a tree's ability to perform all its functions becomes more and more evident. In addition, since animals, including humans, cannot produce their own food and since most of the food we consume is either directly or indirectly of plant origin, it becomes apparent that plants are fundamental to our very survival.

In addition, understanding that oxygen is critical to the survival of roots and therefore, the entire tree, requires that we look at trees in the context in which they grow. Trees, like all living things, do not exist in a void. Most of the world's trees are forest trees, part of a forest ecosystem – an association of flora, funga and fauna bound together in a community. A forest tree's closest associate is the soil in which it grows. The importance of soil to a tree's ability to function necessitates a slight detour from our primary character to review some soil basics:

- soil is crucial to all life on land, as Charles Kellog so eloquently stated in 1938: "Essentially, all life depends upon the soil … There can be no life without soil and no soil without life; they have evolved together." and to quote from the Global Soil Biodiversity Initiative: "Soil biodiversity is largely unseen but is crucial for healthy ecosystems, and ultimately healthy humans." (https://www.globalsoilbiodiversity.org)
- soil is a product of mutualistic relationships such as lichen and mycorrhizae, both of which have facilitated the building of soil structure and the evolution of land plants and their varied communities;

- soil can take millennia to evolve to the point where it can support a forest community;
- soil texture and structure in addition to parent material and climate are the primary determinants of what plant growth can be supported; soil texture is the physical texture of soil, based on the size, shape and percentage of its main constituent particles: sand, silt and clay. The percentage of each of these particles affects such things as water-holding capacity and the soil's permeability (the ease with which air and water can pass through);
- soil structure is the aggregation of sand, silt, clay as well as organic matter particles into clusters; the stability of those aggregates – their ability to remain intact during rainfall and the movement of water through the soil – is determined by soil texture percentages;
- soil aggregation is also directly related to the formation of soil pores inside as well as outside the aggregates; micropores, normally found between the particles inside the aggregates, provide water retention, facilitating root penetration and growth, while the macropores found between the aggregates provide aeration in the soil, critical to root growth and to varied processes in the tree.

Tree Form and Architecture

The generic tree as pictorial mystical metaphor has an upright form with its roots in the earth and its branches reaching for the heavens. Just as any examination of *Homo sapiens* will reflect our incredible diversity, a closer look at the more than 60,000 examples of tree species will paint a much more diverse picture.

The earliest example of vascular plants dates back approximately 420 million years. The evolution of the form that we recognize as a tree extended over the next 360 million years to the appearance of Gymnosperms bearing naked seeds. The most well known remnant from this first recognizable tree form is *Ginkgo biloba*, the Maidenhair tree, followed by the Conifers' original appearance in the Carboniferous period. Angiosperms made a well-defined appearance starting in the Cretaceous period and, joining the Gymnosperms, represent the dominant tree groups with us today.

A great deal of genetic tree architecture evolved in forests. Most of today's tree species originate in forests where many different tree species coexist. Forest trees live in an extended cooperative and connected community below ground, sharing resources and critical environmental information through mycorrhizal associations. Within that context of forest community, trees are competing for access to light, primarily through species' varied approach to light capture – how different tree species arrange their leaves. In 1970, Hallé and Oldeman, in *Tropical Trees and Forests: An Architectural* Analysis (Springer-Verlag, Berlin. 1978), presented a species-specific architectural model, establishing an "inherent growth strategy that defined both the manner in which the plant elaborated its development and the resulting architecture".

Trees come in an assortment of shapes and sizes with an extensive variety of branch arrangement, leaf, flower, fruit and seed shapes and colors as well as a vast array of bark patterns, color and structure – so diverse and artistic as to inspire

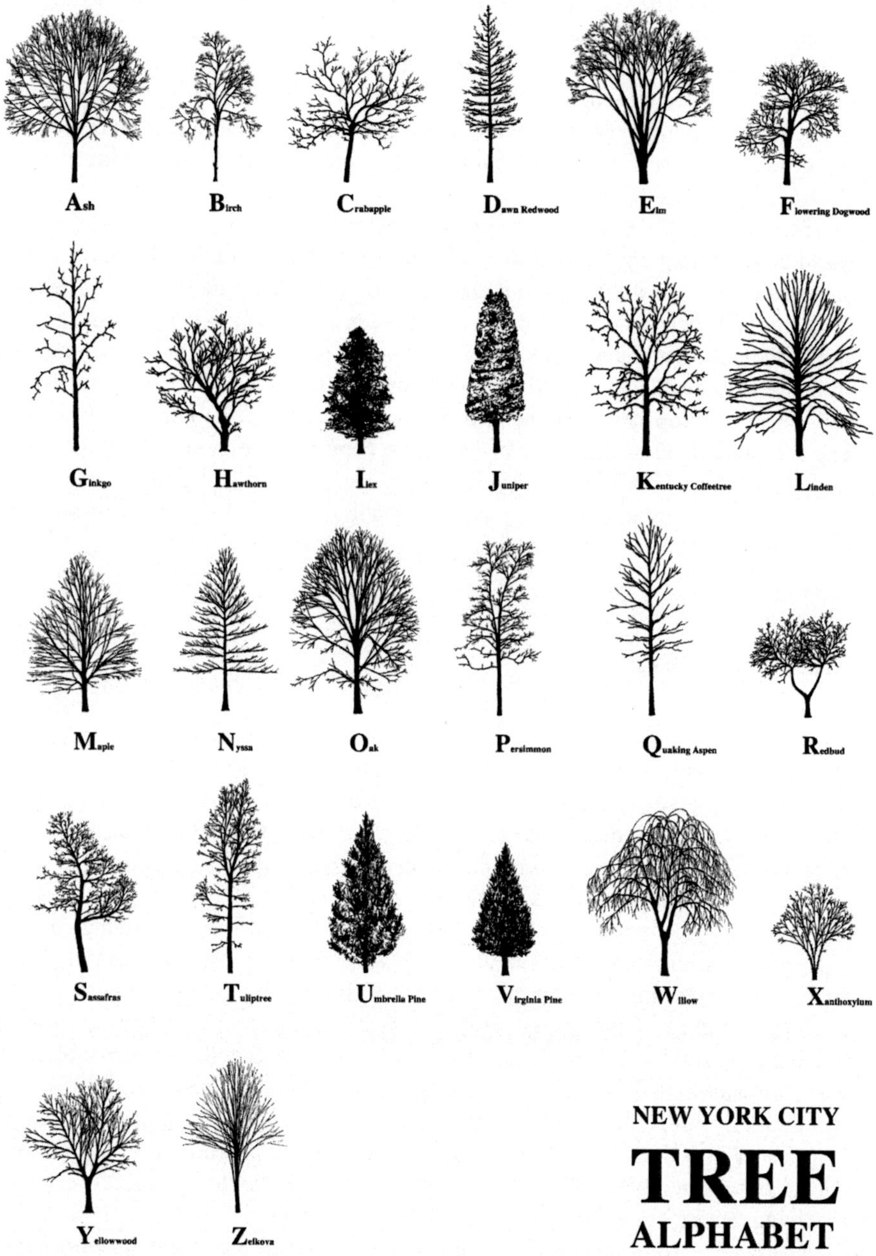

Fig. 1.4 New York City Tree Alphabet. Katie Holten

Cédric Pollet's photo essay *Bark: An Intimate Look at the World's Trees*. The overall crown architecture of each and every species can determine how light is captured and distributed to all the tree's photosynthesizing parts which vary, in turn, depending on the surrounding environment's effect on all those parts and their independent impetus to grow. There is also crown architecture variation within species depending on age – juvenile or mature.

In addition, as the energy that drives a tree's systems is a precious and limited resource, the forest tree, in managing itself, must be extremely frugal in how it allocates that resource. Branches that are no longer producing more than they are using are allowed to die and to be shed but not before any and all stored resources are re-absorbed back into the tree.

Trees can alter their species-specific architecture by varying their growth responses based on prevailing environmental conditions or stress, e.g. smaller leaves or putting on wood exactly where it is needed in response to the stress the tree is experiencing. Many of the factors that have influenced the extensive variation in branching patterns and overall crown architecture, e.g. gravity, wind or apical control and / or dominance, are extremely detailed and beyond our focus but an awareness of these factors can help inform appropriate planning and management, as we shall see going forward. The silhouette models for many of our more common tree genera, as illustrated in Katie Holten's 2018 *New York City Tree Alphabet*, are an excellent approximation of the extensive variety in branching pattern possibilities. (Fig. 1.4 – (https://www.katieholten.com/public-realm/#/new-york-city-tree-alphabet/)

Measurable Structure

Structure is another aspect of similarities and differences between trees and people. We both use living and dead cells and tissues structurally. In humans, hair, nails and the outermost layer of skin are comprised of dead cells, derived from one of the body's organ systems that serves to separate the outside from the inside. These dead cells and tissues aid in regulating internal temperatures as well as forming a barrier between the outside world and living cells and tissues.

As with people, trees' non-living parts are inextricably linked to living cells and tissues in a continuous cycle of creation, growth, decline and death. As a tree evolves from seed to seedling, sapling, juvenile and then maturity, the percentage of living as compared to non-living cells, steadily decreases. A tree's living cells are concentrated in its buds, roots, leaves, the cell layers of inner bark called cork cambium and vascular cambium, creating lateral growth as well as the meristematic cells at the tips of roots and shoots – apical growth. The rest and greatest portion of the tree consists of non-living wood cells that provide support, structure and protection. Although technically non-living, these cells continue to be the transport mechanism for living matter, essential to a tree's ability to be a tree.

Assessing a tree's measurable structure can provide a great deal of information about a tree's condition as well as the quantifiable Ecosystem Services it has the ability to deliver. Trees, especially mature canopy trees, have the ability to deliver an extraordinary array of benefits to the communities in which they exist. These benefits – the capacity to trap and remove air pollutants, store and sequester carbon, improve water quality through filtration and infiltration, reduce storm water runoff through interception, reduce energy consumption and the urban heat island effect through shading and evapotranspiration – are the services an ecosystem or its parts, in this case trees, contribute to human well-being. Information detailing the above ground portion of the tree, e.g. total height and crown dimensions such as live crown or percent of dieback, can provide a current snapshot as well as form the basis for a forecast for ecosystem services deliverables. It is possible to quantify these structural measurements, using valuation methodology software, to determine the volume of Ecosystem Services produced as well as the local financial value of those delivered benefits.

It is critical to understand that a species' capacity – what a tree genetically has – its potential – and the individual's ability to deliver benefits – what a tree can dynamically do with it's genetic capacity – is primarily driven by the tree's canopy – its living crown. The crown, however, does not exist in a void and can only provide those essential benefits if the entire tree organism has vitality – a tree's dynamic ability to grow, reproduce and adapt to its surroundings which can be increased/ enhanced by providing trees with their essential needs and reducing their stress load.

What Do Trees Symbolize – The Tangible and the Intangible Tree

Approaching a tree we approach a sacred being
who can teach us about love and about endless giving.
She is one of millions of beings who provide our air, our homes, our fuel, our books.
Working with the spirit of the tree can bring us renewed energy,
powerful inspiration, deep communion.
– Druid Tree Lore

Trees have facilitated our very existence from the time of our arrival on the evolutionary scene. Throughout human history, trees have enabled us in a very tangible way, bestowing their largesse on us. We live in trees – they surround us with building materials and furniture. We ingest trees – they sustain us with food and medicine. We share our thoughts and our knowledge using trees – we write on paper made from their pulp. We wear trees – they outfit us with fiber and fabric for clothing. Trees convey us via canoes, wagons, bridges. They keep us warm in winter with the fuel source they provide and they keep us cool in summer with their shade. But these same tangibles have also inspired the intangibles – shelter can also be viewed as protection and sustenance can be of the spirit and the mind as well as for the body.

From the beginning of human time, trees as symbols and animators, have nurtured and empowered our spirit by capturing our collective imagination in myth and folklore. If one examines the plethora of tree-based mythology, one can find a common thread – an all-pervading sense of a lofty presence worthy of human aspiration.

The qualities that myth has attributed to trees – venerable, wise, strong, noble, powerful – lend themselves to the underlying symbolism, as noted by Nathaniel Altman in his wonderful book *Sacred Trees*: *Spirituality, Wisdom & Well-Being*. (New York NY. Sterling Publishing, 2000). Altman has stated "Many cultures believed in a Cosmic Tree (Fig. 1.5): a great mythical tree whose roots often penetrated the underworld while its branches reached far into the heavens, a tree that formed the center or axis of the world."

Fig. 1.5 Gaia Mother Tree. Ernesto Neto 2018. Naomi Zürcher

Altman divided the sacred or Cosmic tree into three main groupings:

- the World Tree, as a representation of the universe and the connection of the divine with all living beings,
- the Tree of Life, reflecting a more personal view of the positive relationship between beings and the environment they inhabited,
- the Tree of Knowledge, similar to the Tree of Life but often from a more negative portent as in the biblical *Tree of Knowledge of Good and Evil.*

According to Altman, what makes a tree sacred is "recognition of the power it expresses… Sacred trees have been found to play a part in community life in nearly every culture" (Fig. 1.6).

While many early cultures had creation myths with the Cosmic Tree as an imaginary, conceptual tree that embodied all the beneficial attributes, other tree-oriented creation/ end-of-life stories and cultural lore used tree Genera that actually existed and still exist today.

Fig. 1.6 Haida Totem with Trees. Vancouver 2016. Naomi Zürcher

Our global tree journey takes us to each of the continents to visualize just how the tree →← people connection was ever-present as well known Genera (Fig. 1.7 – North America, South America, Antarctica; Fig. 1.8 – Europe, Asia, Oceania, Antarctica):

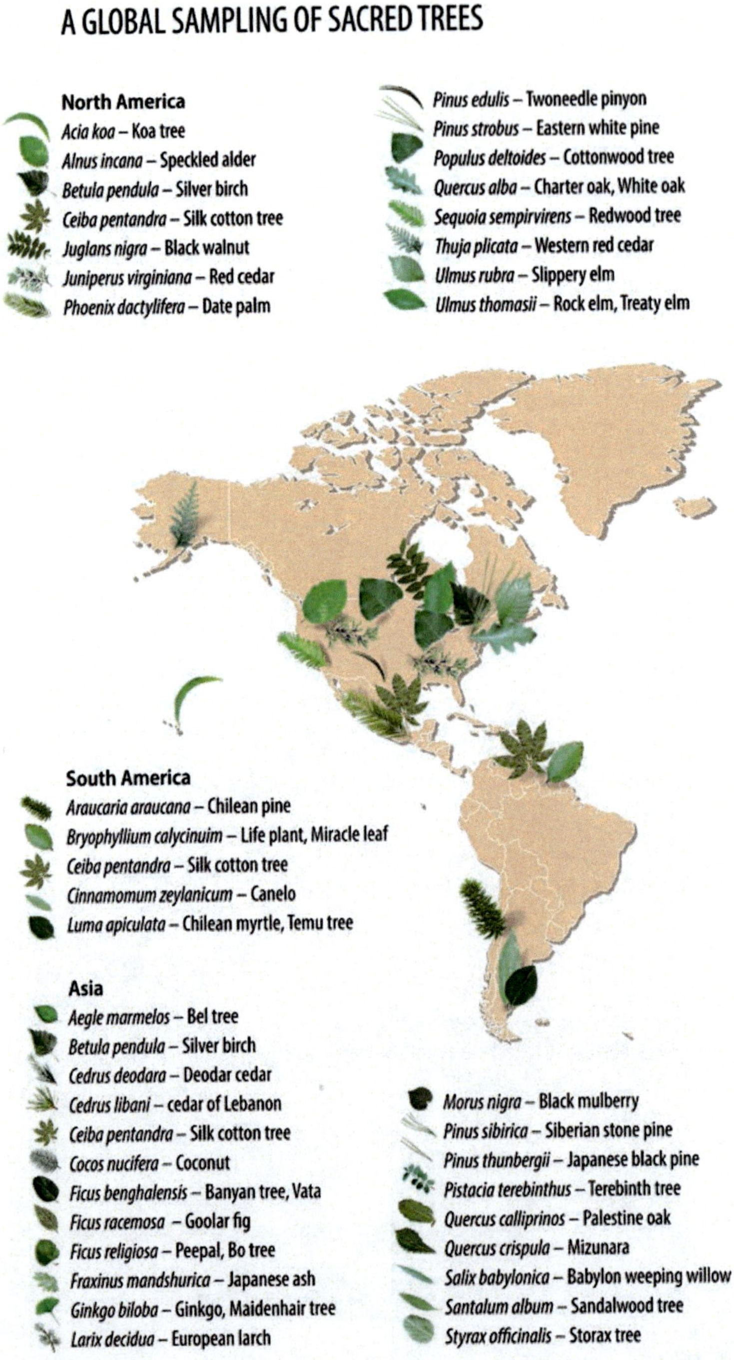

Fig. 1.7 North America, South America, Antarctica. Kathrin Blümli

Europe
- *Abies alba* – Silver fir
- *Alnus glutinosa* – Black alder
- *Betula pendula* – Silver birch
- *Corylus avellana* – Hazelnut
- *Cydonia vulgaris* – Quince
- *Fagus sylvatica* – European beech
- *Fraxinus excelsior* – European ash
- *Fraxinus ornus* – Flowering ash
- *Juglans regia* – Persian walnut
- *Laurus nobilis* – Laurel tree
- *Malus domestica* – Apple tree
- *Morus nigra* – Black mulberry
- *Pinus sylvestris* – Scots pine
- *Platanus orientalis* – Oriental plane tree
- *Pyrus communis* – Pear tree
- *Quercus robur* – European oak, Chapel oak
- *Salix alba* – White willow
- *Sorbus aucuparia* – European mountain-ash
- *Styrax officinalis* – Storax tree
- *Taxus baccata* – yew
- *Tilia cordata* – linden, Littleleaf linden

Africa
- *Adansonia digitalia* – Baobab tree
- *Ceiba pentandra* – Silk cotton tree
- *Celtis australis* – Meis tree
- *Cola nitida* – Kola tree
- *Cordia sinensis* – Nyuot tree
- *Ficus sycomorus* – Sycamore fig
- *Morus nigra* – Black mulberry
- *Pseudolachnostylis maprouneifolia* – Musoli tree
- *Tamarindica indica* – Tamarind tree

Oceania
- *Agathus australis* – Kauri tree
- *Pandanus tectorius* – Pandanus tree

Fig. 1.8 Europe, Asia, Oceania, Antarctica. Kathrin Blümli

The imagery of the continents covered in leaf symbols is merely a sampling of the sacred trees, venerated in creation stories, tree spirit legends, fertility rituals and celebrations of cycles – seasons, birth and death, planting and harvest – and gives one a sense of the vast extent to which trees have informed every aspect of life from our beginnings into the present.

"The leaves of the trees were for the healing of the nations"
Revelation 22:2

A profound example of the extraordinary relevance today of cultural tree lore of yesteryear can be found in the legend of the Great Tree of Peace. Almost a thousand years ago, the Nations known today as the Haudenosaunee (Iroquois Confederacy) were at war with each other. Skennenrahawi (Two River Currents Flowing Together – who came to be known as the Peacemaker) had a vision of the Tree of the Great Long Leaves (the Great White Pine – *Pinus strobus*) reaching to the sky. Beneath the pine, the weapons of war were buried and four white roots extended in all four cardinal directions to the corners of the Earth. The name of these roots was The Great White Roots and their nature – Peace and Strength. Those who followed the roots to the base of the Great White Pine, found shelter beneath its branches. At the top, an eagle was perched, alerting the People of approaching danger.

Skennenrahawi, with the assistance of Hayonhwatha, visited each of the Haudenosaunee nations – Mohawk, Seneca, Oneida, Cayuga and Onondaga – in an effort to realize his vision of Peace. Although rejecting the vision at first, the Five Nations came together in the shade of the Tree of the Great Long Leaves – the Great Tree of Peace (Fig. 1.9 – symbolized by the Hayonhwatha wampum belt http://www.onondaganation.org/culture/wampum/hiawatha-belt/) – and formed the Council of the Great Peace, Thus, this League of Five Nations (now six with the later addition of the Tuscarora) crafted Kaianerekowa, *the Great Binding Law of Peace*, under which all could live together in Peace, as they do to this very day – peace, not in terms of the absence of war but rather, as the fundamental way of life and as embodied in the Law. (Kaianerekowa, The Great Binding Law, was the first of its kind on the American continent. Benjamin Franklin and the framers of the US Constitution borrowed extensively from the Great Binding Law in formulating the American model.)

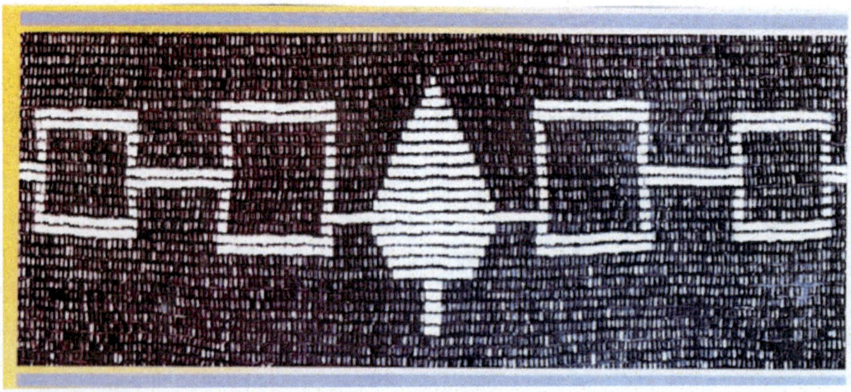

Fig. 1.9 Haudenosaunee Great Tree of Peace Hayonhwatha Belt

The legacy of the Great Tree of Peace has bestowed its wisdom on more than just the Constitution of the United States. The White pine – *Pinus strobus* – and its symbolism of a government rooted in the natural world, came to be known as The Liberty Tree, displayed on colonial flags. Eagle-that-sees-far – the Eagle protector, perched atop the Tree of the Great Long Leaves – became the American Eagle. The bundle of 13 arrows clutched by the American Eagle on the US Great Seal had its origins in Skennenrahawi's bundle of 5 arrows, symbolizing the strength through unity of the Five Nations.

In 1984, Chief Jake Swamp, of the Wolf Clan Mohawk, established the Tree of Peace Society. On September 17, 1987, the Tree of Peace Society planted a Tree of Peace at the Washington D.C. Mall. At this tree planting ceremony, led by Leon Shenandoah, Chief of the Iroquois Confederacy and Chief Oren Lyons, Onondaga Nation, Chief Jake offered the following sentiments as he placed an arrowhead, symbol of the weapons of war, in the planting hole: "If we have feelings of war among us, we must take them out and bury them. If we have feelings of greed, we must take them out and bury them. Feelings of mistrust must be buried so we can start to create the atmosphere of peace for future generations". (Paul A. W. Wallace, *White Roots of Peace: The Iroquois Book of Life.*^ Clear Light Publishers, 1994.) Chief Jake, in his mission to bring to the world the teachings of Skennenrahawi and to advocate for greater ecological awareness and sensitivity, planted hundreds of Trees of Peace in all corners of the world. I had the privilege of meeting him at the Cathedral of St. John the Divine (NYC) Tree of Peace planting in the early 1990s.

The forests are the flags of nature.
They appeal to all and awaken inspiring universal feelings.
Enter the forest and the boundaries of nations are forgotten.
It may be that some time an immortal pine
will be the flag of a united peaceful world.
Enos A. Mills

On a personal level, I believe it may be hard to separate the imaginary or even the specific tree from the tree with personal relevance – the Cosmic tree from a personal mystical experience, the tree on canvas from the lofty tree in our own mind's eye. We revere trees we have assigned Heritage status to – monarch trees that have lived for centuries – witness-bearing trees that embody the memory of millennia. One might also wonder if the ever increasing willingness to defend not just the environment, but trees and forests in particular, is rooted in our ancestral connections to these sentient spirits: loss of trees and forests is a loss of soul mate. It is clear that trees animate people, animation that has given impetus to the vast array of tree-oriented creativity in the arts.

Painters and sculptors have explored tree imagery in depth, from the literal to the abstract, capturing the tree through their own visualization processes. Just prior to the turn of the twentieth century, French Impressionist Claude Monet used trees extensively in his paintings, as exemplified in his 1891 Poplar series. With the turn of the twentieth century, Impressionism gave way to early modern art in the form of Fauvism and early Cubism. The Fauvists took their initiative from the post impressionist approach to nature and the likes of Vincent Van Gogh with his imagery of peach, apricot, almond and olive trees and, most importantly, from Paul Cézanne with his wonderful visions of L'Estaque, a village just east of Marseilles France. In these painted landscapes, trees often frame the exquisite vistas as in *Vue sur L'Estaque et Le Château d'If*. These images from the Father of modern art inspired a visit from Georges Braque shortly after Cezanne's death in 1906 and the painting *Paysage à L'Estaque*. That visit planted the seeds of Cubism which blossomed upon Braque's return in the 1907 version of *Paysage à L'Estaque*.

The Dutch painter Piet Mondrian depicted his artistic evolution in his well known Tree Series in which, over a period of 6 years, his tree paintings evolved from the realistic *Evening Red Tree*, painted in 1908 to a more Theosophist view, beginning in 1912 with *The Grey Tree* and *The Flowering Apple Tree*. Mondrian's final tree painting *Arbre*, painted in 1913, reflected his movement to analytical Cubism and beyond. Regardless of how they were rendered – realistically or abstract – or the artist's intended focus, Mondrian's trees continue to embody both the energy and the ethereal quality ever present in the living entity.

Our wanderings through treed visual arts wouldn't be complete without including two unique sculptures. The first, Alberto Giacometti's 1950 bronze *The Forest* (Fig. 1.10 © Succession Alberto Giacometti / 2018, ProLitteris, Zurich). The elongated figures stand rooted as do trees in a forest; individuals standing seemingly alone but bound together by their lives in community, in association – a metaphor for the energy driving the motionless movement of the very trees being portrayed.

Fig. 1.10 The Forest. 1950. © Succession Alberto Giacometti / 2018, ProLitteris, Zurich

The second, Alexander Calder's extraordinary stabile entitled *The Tree*, currently located in the Landschaftspark adjoining Fondation Beyeler in Riehen Switzerland (Fig. 1.11, Missouri Botanical Garden, Library of Congress Historic American Buildings Survey, Survey number: HABS MO-1135).

Calder's endless fascination with the Tree as form and essence, was re-envisioned over and over again and captured in a Fondation Beyeler 2013–14 exhibition *Alexander Calder: Trees – Naming Abstraction.*

Fig. 1.11 The Tree. 1966. Alexander Calder

Tree imagery has not limited itself to the visual arts. C J Jung, in his essay entitled "The Philosophy of Trees" explored the effect of trees on individuals' imagination in the form of fantasy, mythology and religion. Trees have permeated song such as Rush' *The Trees,* prose and poetry from the likes of William Shakespeare, Robert Frost, William Carlos Williams, Walt Whitman. Many inspirational examples abound from an assortment of pens.

Herman Hesse offered a profound commentary on trees in his *Wanderings*:

> "Trees are sanctuaries. Whoever knows how to speak to them, whoever knows how to listen to them, can learn the truth. They do not preach learning and precepts, they preach, undeterred by particulars, the ancient law of life.
>
> A tree says: A kernel is hidden in me, a spark, a thought, I am life from eternal life. The attempt and the risk that the eternal mother took with me is unique, unique the form and veins of my skin, unique the smallest play of leaves in my branches and the smallest scar on my bark. I was made to form and reveal the eternal in my smallest special detail." (Herman Hesse. *Wandering: Notes and sketches.* London UK: Triad/Panther Books. 1985, 49.)

And in his poem *Woods,* Wendell Berry shapes an intimate woodland experience in which his own senses take on a tree persona (Fig. 1.12)

Fig. 1.12 Old Growth *Thuja plicata* along Siigee (McIntyre Bay), Haida Gwaii. Naomi Zürcher 2016

I part the out thrusting branches
and come in beneath
the blessed and the blessing trees.
Though I am silent
there is singing around me.
Though I am dark
there is vision around me.
Though I am heavy
there is flight around me.
Wendell Berry
Woods

Native American story and song are filled with earth images in which trees play an important role. Black Elk, in his vision of the Sacred Circle, framed the importance of the tree to the People of the Sacred Hoop:

"In the old days, all our power came to us from the sacred hoop of the nation and so long as the hoop was unbroken the people flourished. The flowering tree was the living center of the hoop, and the circle of the four quarters nourished it." (Black Elk, as told to John G. Neihardt (Flaming Rainbow). *Black Elk Speaks: Being the Life Story of a Holy Man of the Oglala Sioux.* New York NY: William Morrow Press. 1932. Chapter XVII. The First Cure.)

From my own perspective, one of the greatest tributes to the tree not only as balm for the human soul but as healer of the earth was expressed in Jean Giono's *The Man Who Planted Trees* – http://www.pinetum.org/GionoUK.htm. In this allegorical tale, Elzéard Bouffier singlehandedly turns a valley that is a wasteland into a lush forested paradise by planting thousands of oak, beech and birch trees over four decades.

The oaks of 1910 were now ten years old and were taller than me and than him. The spectacle was impressive. I was literally speechless and, as he didn't speak himself, we passed the whole day in silence, walking through his forest.

It was in three sections, eleven kilometers long overall and, at its widest point, three kilometers wide. When I considered that this had all sprung from the hands and from the soul of this one man – without technical aids –, it struck me that men could be as effective as God in domains other than destruction….

In the direction from which we had come, the slopes were covered with trees, twenty to twenty-five feet tall. I remembered how the land had looked in 1913 – desert… Peaceful, regular toil, the vigorous mountain air, frugality and, above all, serenity of spirit has endowed this old man with awe-inspiring health. He was one of God's athletes. I wondered how many more acres he was going to cover with trees….

He had followed his idea, and the beeches that reached up to my shoulders and extending as far as the eye could see bore witness to it. The oaks were now good and thick, and had passed the age where they were at the mercy of rodents; as for the designs of Providence, to destroy the work that had been created would henceforth require a cyclone. He showed me admirable stands of birches that dated from five years ago, that is to say from 1915, when I had been fighting at Verdun. He had planted them in the valley bottoms where he had suspected, correctly, that there was water close to the surface. They were as tender as young girls, and very determined. This creation had the air, moreover, of working by a chain reaction. He had not troubled about it; he went on obstinately with his simple task. But, in going back down to the village, I saw water running in streams that, within living memory, had always been dry. It was the most striking revival that he had shown me. These streams had borne water before, in ancient days. Certain of the sad villages that I spoke of at the beginning of my account had been built on the sites of ancient Gallo-Roman villages, of which there still remained traces; archeologists digging there had found fishhooks in places where in more recent times cisterns were required in order to have a little water.

The wind had also been at work, dispersing certain seeds. As the water reappeared, so too did willows, osiers, meadows, gardens, flowers, and a certain reason to live….

When I consider that a single man, relying only on his own simple physical and moral resources, was able to transform a desert into this land of Canaan, I am convinced that despite everything, the human condition is truly admirable. But when I take into account the constancy, the greatness of soul, and the selfless dedication that was needed to bring about

> this transformation, I am filled with an immense respect for this old, uncultured peasant who knew how to bring about a work worthy of God.
>
> Elzéard Bouffier died peacefully in 1947 at the hospice in Banon.

If one analyzes even this small sampling of the myths, legends, visual and written art, one can see the effect trees have always had on our psyche, individually as well as collectively. As more and more of us take up residence in urban centers, it would seem that we create more and more distance between ourselves and what has always served as sustenance for mind, body and spirit. Given what our human dynamism has wrought environmentally, how then do we bring those invaluable ancestral connections into an urban ecosystem that can support and sustain life? Enter the Urban Forest.

Chapter 2
The Urban Forest: From Oxymoron to Critical Resource

What is the Urban Forest; how does the philosophical underpinning of inclusiveness contribute to its relevance and what makes it such a critical resource in today's world?

Keywords Climate change · Greenhouse gas emissions · Human needs · Nature-based · Philosophy · The public · Sustainability · Urban forest · Urban heat island · Water resource management

Fig. 2.1 Phiedra by PEJAC I Spain, Madrid I 2013. Photo credit- @pejac_art

Trees are features in landscapes which extend far beyond property lines.
They belong in a sense to the entire population.
They are our shade trees rather than exclusively your or my shade trees.
They are heritages of a succeeding generation, as well as our own.
They are material assets of private property, and at the same time
contribute substantially to the well-being of the community,
the state and the nation.
Ephraim Porter Felt, 1938

N. Zürcher, *Connecting Trees with People*, Future City 16,
https://doi.org/10.1007/978-3-030-94534-3_2

The Urban Forest as Philosophical Oxymoron

The term Urban Forest has been defined in Wikipedia as "a forest or a collection of trees that grow within a city, town or a suburb…" (28 August 2019) This definition may have sufficed when the term Urban Forest first arrived on the scene but, in today's reality, the Urban Forest is much more than that. Dr. David Nowak, USDA Forest Service Research Scientist, offered the following simple and expanded definition of urban forests: "Urban Forests are ecosystems characterized by the presence of trees and other vegetation in association with people and their developments" (Nowak et al. 2001). I have taken the liberty of broadening the *ecosystem* aspect in David's definition with his knowledge and consent:

> *Urban Forests are ecosystems characterized by the presence of trees and related flora, funga and fauna, the soils and landscapes they populate and the air and water resource they coexist with, all in a dynamic association with people and their human settlements.* (Naomi Zürcher, 2018)

If we follow the evolution of the Urban Forest from pre-urbanized land use to present day urban development, we can find evidence of common woodlot holdings dating back to the seventeenth century. This *land in common* approach to holdings of trees in and around human settlements reinforces the millennia-old understanding of just how essential trees were to people. Beyond the myth and legends, that early understanding only went so far as to serve the community's needs for building materials, food, clothing – the tangibles.

The beginnings of the Urban Forest of today as a philosophical approach to urban living had its origins in the tree as amenity. Such thinking did not take root until the nineteenth century with the likes of Frederick Law Olmstead who believed that "trees and vegetation enhanced the morale and counteracted the anxieties of city life. A city park was to provide a natural verdant and sylvan scenery for the refreshment of town-strained, men, women, and children." (Forest History Society https://foresthistory.org/education/trees-talk-curriculum/trees-in-your-own-backyard/essay-history-of-urban-forests/) The Urban Forest has evolved with the understanding of the need that Olmstead clearly embraced in his concept and his designs – the need for nature in places where nature no longer has a presence.

When the actual term Urban Forest first came into usage, it was greeted as one greets any oxymoron – mostly with a *What's that???* – and, not surprisingly. What an odd combination of words. If one pictures a forest, connecting that to the word *urban* is an anomaly to everything a forest represents. It's only when we give up traditional concepts and start to analyze the world as it presently is – the one we are actually living in – that such an oxymoron begins to resonate and takes on clarity.

A further development of Urban Forest from a philosophical perspective requires an exploration of terms such as ecosystems and sustainability. Everything exists within a web of relationships, in other words, an ecosystem. For built environment dwellers, it's an urban ecosystem. What is critical in sustaining an ecosystem, urban or otherwise, is that our interventions and manipulations do not impede, restrict or eliminate the availability of the ecosystem services we and all other living things depend on. More importantly, that we do not destroy the very resources that deliver those ecosystem services as well as a multitude of other essential benefits, tangible and intangible.

In 1987, the Brundtland Commission (formerly referred to as World Life Issues on Environment and Development) defined sustainable forestry as "managing our forests to meet the needs of the present without compromising the ability of future generations to meet their own needs…"

Defining sustainability with a primary focus on the *needs* of the recipient continues the *all about us* one-sided approach, but if one also looks at sustainability philosophically, the focus is on natural systems and how they function. The sustaining of an ecosystem is not a one-sided affair. A sacred and reciprocal approach to land and its connection to people was very much in evidence in the larger world view of the Haudenosaunee's Great Binding Law of Peace referenced in Chap. 1.

The purpose and the action of sustaining something cannot just be about us. The idea that we are the only entity that requires sustaining is part of the reason we are where we are today. The idea that all the Earth's resources are there for the taking results in life out of balance (Koyaanisqatsi [Hopi] – Life out of Balance. Godfrey Regio, director; Philip Glass, music composition. 1982). The earth – SOIL – cannot sustain us if we do not sustain it; trees cannot sustain us if we do not sustain them, not just because of what these organisms do for us but because of the organisms themselves – living organisms that are, in Haudenosaunee terms, an inseparable part of sacred landscapes or in western vernacular, part of an ecosystem. We are also part of that ecosystem, not separate from it and, as such, what we do to whatever we are part of, we do to ourselves along with everything else within the system.

This we know:
the earth does not belong to man – man belongs to the earth…
All things are connected like the blood which unites one family…
Whatever befalls the earth, befalls the sons of the earth.
Man did not weave the web of life – he is merely a strand in it.
Whatever he does to the web, he does to himself.
Chief Seattle in a letter to Pres. Franklin Pierce, 1855

In addition, much of our urban ecosystem *green ways* planning tends to be curvilinear or linear in design and construct, guiding us along *greened* transportation corridors, a manufactured construct transporting us to something *out there*, outside of our community, outside our circle, beyond our intimate connections.

Rather, we might look at our ancestors or an aboriginal perspective, which was circular and cyclic, reinforcing the concept of the web of life and its interconnected relationships, promoting balance and a holistic harmony with the natural world – an ecosystem approach to existence. (Figure 2.2 Haudenosaunee Confederacy Tree of Peace logo).

Fig. 2.2 Haudenosaunee Confederacy logo. Haudenosaunee Confederacy

Within a nature-based sustainability model, there is also a keen belief in both adaptability and change, but change that further promotes balance and harmony, not change that initiates distress, death and the depletion of the Earth's populations and resources – a circular process in which we are part of the world around us – part of that ecosystem (Fig. 2.3 First Nations Pedagogy Social Ecological model).

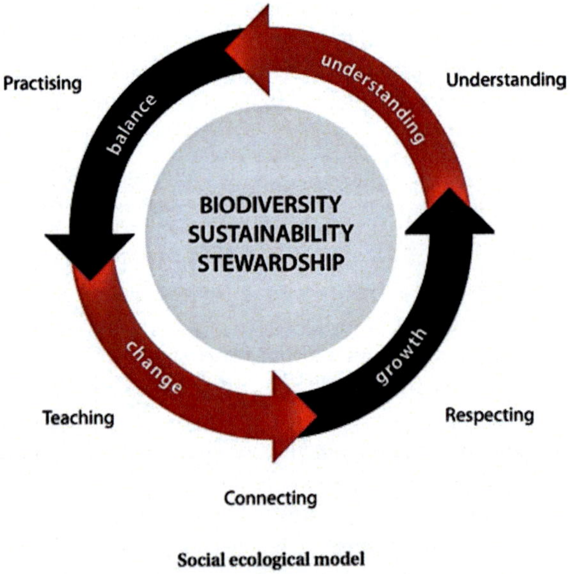

Fig. 2.3 Social Ecological Model Credit First Nations Pedagogy

As history and legend have shown us, trees and the landscapes they populate were, and continue to be, traditionally revered by Indigenous peoples. That is not to say that our First Peoples did not use the resource but it was used in a sacred way with respect for the sentience of the resource, with reciprocity and in balance with what was available. Some of our present day difficulties can be laid at the door of the Judeo-Christian biblical injunction "to fill the earth and subdue it…". In addition, the European ethic established that the forest belonged to those who brought it under the plough. If one combines that with the fact that land value was determined by and equated with the value of the commodities that could be extracted from it, our present state of affairs is not at all surprising. This type of land ideology did much to instigate our westernized view of the Earth and to undermine the integrity of the Earth and all its constituent parts.

The population shift from rural landed life to urban has further exacerbated this dilemma. While the urban environment is an entirely man-made construct and almost everything in it is UN-natural and artificial in a *construct* sense, I believe the need to have trees, their flora, funga and fauna associates and the landscapes they populate within that man-made construct, speaks to our fundamental human needs. Early towns did not have such a need for green amenities – *wild nature* was just outside the town limits – easily seen and readily accessed. As towns grew into cities and people became more and more physically removed from *wild nature*, a psychological void developed (Kardan et al. 2015), attesting to the irrefutable fact that we do need nature because we are part of nature, irrespective of biblical injunctions and not just for those tangible products but for the *entirety*.

Oxford online dictionary (https://www.lexico.com/definition/entirety n.d.) defines entirety as *the whole of something*. If, as stated earlier, we are actually living in a dynamic association with all other components of the urban ecosystem, we are very much a part of the *entirety* and as being part, we are INTER-connected with all of it. No matter the culture, we all have an agrarian heritage exemplified by the breadth of the mythology and folklore explored in Chap. 1. While we all have our historical roots in the land, our consolidation with the built environment has severed those roots and many of us have lost those connections. Re-establishing our connection to the land, to trees – our trees – is, in effect, a way of finding our way back to our roots – of finding our way back into the circle, the web of life.

Why is that so critical in today's world? The plethora of reports and environmental analyses present an overwhelming body of knowledge as to the dismal state of our environmental affairs.

On 20 November 2018, the World Meteorological Organization reported that greenhouse gases are continuing to rise at unprecedented levels. "CO_2 remains in the atmosphere for hundreds of years and in the oceans for even longer. There is currently no magic wand to remove all the excess CO_2 from the atmosphere," said WMO Deputy Secretary-General Elena Manaenkova.

"Every fraction of a degree of global warming matters, and so does every part per million of greenhouse gases," she said. https://public.wmo.int/en/media/press-release/greenhouse-gas-levels-atmosphere-reach-new-record

Also in November 2018, the U.S. Global Change Research Program issued its Fourth National Climate Assessment report. In the report's Overview, the introductory paragraph states "Earth's climate is now changing faster than at any point in the history of modern civilization, primarily as a result of human activities. The impacts of global climate change are already being felt in the United States and are projected to intensify in the future—but the severity of future impacts will depend largely on actions taken to reduce greenhouse gas emissions and to adapt to the changes that will occur…"

And the 2019 Global Assessment, published by the Intergovernmental Science-Policy Platform on Biodiversity and Ecosystem Services (IPBES), states emphatically and unequivocally indisputable Climate Change-related facts such as "Nature and its vital contributions to people, which together embody biodiversity and ecosystem functions and services, are deteriorating worldwide... Human actions threaten more species with global extinction now than ever before." "Direct and indirect drivers of change have accelerated during the past 50 years. The rate of global change in nature during the past 50 years is unprecedented in human history." And "Goals for conserving and sustainably using nature and achieving sustainability cannot be met by current trajectories, and goals for 2030 and beyond may only be achieved through transformative changes across economic, social, political and technological factors." https://www.ipbes.net/sites/default/files/downloads/spm_unedited_advance_for_posting_htn.pdf

With each report, extreme weather event, severe air pollution day, etc., etc., the urgency becomes more and more clear. While it is apparent that we must directly reduce consumption of whatever produces greenhouse gas emissions, we must also use every tool and every strategy to mitigate the problems we have created and adapt to what is already on our collective doorsteps as well as on the horizon.

As the globe continues to urbanize and cities relentlessly expand, the vast array of environmental negatives produced by our all-consuming existence increases exponentially. For those of us residing in urban centers – the belly of the beast – it is essential that we address what our lifestyle is doing to each and every one of us as well as its impact on everything that we coexist with, both near and far.

So, how can connection-oriented philosophies and a resonating oxymoron be embraced by those of us residing in towns and cities and result in actionable endeavors? In those early *Urban Forest as oxymoron* days, the USDA Forest Service's Urban & Community Forestry Program *ReLeaf* put Urban and Community Forestry partially in the hands of the concerned and interested Public – the grassroots. As a co-author of the NY State ReLeaf program and having implemented the program in NY City for close to a decade, I can personally attest to the groundswell that accompanied the program's realization. That fertile grassroots ground was soil you could plant Urban Forest seeds in and that's exactly what the ReLeaf program did, turning that Urban Forest oxymoron into vernacular. And, Urban Forestry, the research and management aspect which sustains the structure and health of the resource, evolved with the growing awareness of our need for nature in those places and spaces where nature no longer had a viable presence.

Formulaic methods to establish the amenity value of a tree that was too large to be replaced with an equivalent tree date back to the early 1900s. (Watson, G. November 2001) and have continued their development up until today. A whole array of methodologies using varied assessment strategies have been developed to determine such criteria as replacement value, land value, visual aesthetic value, etc.

to establish cost of cure and for use in court proceedings in countries throughout the western world.

During the late twentieth century, scientific research began to explore the tree from another perspective – Ecosystem Services – the environmental benefits a tree can afford urban ecosystem dwellers. That extensive research has shown that mature canopy trees offer one of the most cost-effective strategies for mitigating what ails us. Many regulating Ecosystem Services benefits modify local microclimates to improve living conditions, enhancing both social capital and human habitat.

The urban heat island regulating services that urban trees have the capacity to provide include the following, supporting an Ecosystem Services-based approach to planning and managing the Urban Forest for climate change adaptability:

- solar radiation
- wind speed
- ambient air temperatures
- relative humidity
- re-radiation from paved areas.

Climate change will also see an increase in extreme weather events. Healthy, well-managed urban trees increase urban water-management functions by:

- increasing permeable landscapes
- increasing the bio-filtration of urban water-born pollutants as well as infiltration rates
- recharging ground water
- rainfall interception, reducing runoff volume and speed as well as soil erosion
- runoff interception and storage thus improving the status of both surface and groundwater
- increasing evapotranspiration rates
- increasing water-holding capacity.

In addition, tree canopy and in particular the urban tree canopy, creates a healthier environment by:

- improving air quality
- enabling beneficial levels of gaseous exchange thus improving the greenhouse gas balance
- intercepting polluting particles,

while the presence of a healthy and viable Urban Forest in the heart of cities contributes to

- improved mental health through positive outlook
- a sense of *place* and resulting social bonding in green urban communities
- reduced chronic stress levels resulting in a lower prevalence of anxiety disorders and depression
- improved public health through reduced rates of cardiovascular disease and respiratory-tract illnesses.

Rather than the amenity approach, what Ecosystem Services research revealed was a need to value the tree not only as an amenity but the much more wide-ranging and critically important value of the tree as provider of invaluable Ecosystem Services benefits, which had never before been quantified. With the development and release of valuation methodologies such as the i-Tree Tools in August 2006, trees, especially urban trees, could now be quantified for many of the contributions they were making to human well-being.

Many early studies of Ecosystem Services made the assumption that these services were generated outside cities and were then made available to urban residents. With the development of the i-Tree Eco valuation methodology and its implementation in the US, Canada, Australia, Mexico and now Europe, we can scientifically document the Ecosystem Services attributes that urban trees directly afford urban residents and make available to surrounding communities, providing predictive, dynamic models for Ecosystem Services' quantification and monitoring, facilitating adaptive planning, design and management in tandem with spatial development. The ability to quantify these benefits, not just in volume or percentages but also in local monetary terms, enables cost : benefit comparisons as well as a better understanding of the relationship between tree structure and benefits generated:

> "i-Tree is a state-of-the-art, peer-reviewed software suite from the USDA Forest Service that provides urban and rural forestry analysis and benefits assessment tools. The i-Tree tools can help strengthen forest management and advocacy efforts by quantifying forest structure and the environmental benefits that trees provide." https://www.itreetools.org/about.php

The current version of i-Tree Eco (V. 6) provides for an extensive Ecosystem Services assessment of trees, be they one or many, in forest or Urban Forest. "It is designed to use field data from complete inventories of trees or randomly located plots throughout a community, along with local hourly air pollution and meteorological data, to quantify urban forest structure, environmental effects, and value to communities. Managers or researchers can use the baseline data to make effective resource management decisions, develop policy, and set priorities." i-Tree Eco v6.0 User's Manual https://www.itreetools.org/resources/manuals.php

The here and now ravages of climate change demand that we use all such tools at our disposal to create livable and resilient cities for today and for the future. In addition, professionals must facilitate the incorporation of such tools into participatory citizen programs and citizen science endeavors.

We ought to think that we are one of the leaves of a tree,
and the tree is all humanity.
We cannot live without the others,
or without the tree…
Pablo Casals (1876–1973)

In discussing Climate Change and the plight of our local environment with neighbors near and far, I am often presented with the *What can I do that would really make a difference* question. Readily accessible and free software such as i-Tree can frame participatory endeavors that could include both youth and adults who wish to roll up their sleeves and get involved. It would also provide collaboration possibilities, connecting Government managers and paid professional caregivers with a tree's immediate constituency – its human neighbors. By sharing our expertise and connecting the dots between healthy maturing canopy trees and maximized Ecosystem Services benefits through best management and essential care strategies, we can greatly enhance the tree →← people connection by encouraging communities to improve trees' living conditions and thus improve their own.

Last, the strategic planning for and management of a healthy Urban Forest that will be climate change adaptable requires a comprehensive, multi-tiered plan with a clearly stated vision and goals involving input from an extensive collaboration of stakeholders.

This cadre of actors spans many disciplines from urban planners and ecologists, landscape architects and architects, structural and civil engineers, arborists, urban foresters and foresters as well as an informed Public, all contributing their unique best management expertise and experience to varied management functions.

An informed Public!!! Given that a large part of the Urban Forest is planned and managed using taxpayer funding implies ownership. The Urban Forest is the Public's Forest and, as such, must incorporate the Public as part of a viable management strategy. From my own professional perspective, it is virtually impossible for any governmental authority to manage the resource effectively without the Public. In addition, an informed Public is usually more inclined to offer positive advocacy whenever their Urban Forest is threatened. It is therefore critical to encourage the connections and invite the Public to join in wherever they can be effective.

By means of trees...
wildlife could be conserved, pollution decreased,
and the beauty of many landscapes enhanced.
This is the way, or at least one of the ways,
to spiritual, moral and cultural regeneration.
– E. F. Schumacher

References

Kardan O et al (2015) Neighborhood greenspace and health in a large urban center. Sci Rep 5:11610. https://doi.org/10.1038/srep11610

Nowak DJ, Noble MH, Sisinni SM, Dwyer JF (2001) People & trees: assessing the US urban Forest resource. J For 99:37–42

Watson G (2001) A study of CTL a formula values. J Arboric 27(6):289–297

Chapter 3
The Great Tree Migration: From Forest to City Center

Almost all the trees we plant in our cities come from forests somewhere in the world. What does that mean for the tree? How do our present planning and management strategies affect tree health and resilience?

Keywords Ability · Buttress · Capacity · Ecosystem · Growth · Impervious surfaces · Mycorrhizae · Reiteration · Root system · Soil specification · Woody tissue photosynthesis

Fig. 3.1 Wald. Nils Nova

N. Zürcher, *Connecting Trees with People*, Future City 16,
https://doi.org/10.1007/978-3-030-94534-3_3

So rich and fleecy
were the outlines of the forest
that scarce an opening
could be seen,
the whole visible earth,
from the rounded mountaintop
to the water's edge,
presenting one unvaried hue
of unbroken verdure...
In a word, the hand of man
had never yet defaced or deformed
any part of this native scene...,
a glorious picture
of affluent forest grandeur...
"This is grand! – 'tis solemn! –
'tis an education of itself
to look upon!
exclaimed Deerslayer...
James Fenimore Cooper
Leatherstocking Tales:
The Deerslayer (1841)

While such pristine verdant carpets have not appeared in the literary limelight for many parts of the world, James Fenimore Cooper's The Deerslayer painted a word picture of the experience of the protagonist upon arriving in the New World – endless forest as far as the eye could behold. And Cooper, in his reflection on *the hand of man...*, already bears witness to the effect of man's presence on verdant carpets.

Tree growth, be it in the forest or the streets of a City, is the combined interaction of genetic potential – its capacity – "what you have as a result of your genetic code; a potential source for some future action or product" (Shigo, Dr. Alex L., 2 February 1996) – and its ability to deliver on that capacity, driven by surrounding environmental conditions – "what you are doing with what you have; a dynamic or kinetic process (Shigo, *et al.*).

Historically, almost all of our urban trees originate in a forest somewhere in the world. In traditional forests, the interaction of genetic potential and environmental conditions is fairly stable. In an urbanized area, stability is an anomaly. Given that Urban Forestry has evolved with our need to have nature in places where nature no longer has a presence, it is critical that we re-visit some of what we explored in Chap. 1 and expand on it so we establish a better understanding of the urban tree and take stock of its origins – its forest persona and the context in which it evolved. What exactly is forest or forest ecosystem and how does that drive the genetics that these trees bring to an urban ecosystem (an ecosystem being the sum total of all the living organisms that are part of a specific environment, their interactions with each other and the environment they are part of. In the forest, that includes the climate, the soil and the soil's geologic parent material while in an urban ecosystem it also includes the built environment.)

- A forest ecosystem is a complex adaptive ecosystem – a sophisticated, highly-developed community of trees and all their associates – related flora, funga, fauna and, most importantly, soil with a profile consisting of an organic layer and horizons A (topsoil), B (subsoil) and C (parent material), supporting a mycorrhizal network and containing a functioning soil microbial community, providing air, water and the nutrients, macro and micro, all associates in the forest community depend on, either directly or indirectly.

- Forests are a process of succession, of evolution. Forests don't happen in a day, a week, a year. They occur over millennia, beginning with the specific soils they populate, formed by varied weathering processes of parent material – the rocks and geological sediments specific to the site combined with the forest's detritus that is processed in the A horizon.
- Forests evolve in direct relationship to their environment – temperature range, the soil's texture, structure and pH and the availability of light and water will all dictate what species are growing and where – higher elevations or lower, coastal or inland, temperate or tropical, edge or interior.
- Although a forest may consist of many different species of trees, space above ground is shared in competition while space below ground is shared in community. Trees are social, they are collaborators. They have evolved to depend on a communal rooting landscape – a symbiotic mycorrhizal network providing inter-species communication that enhances self-management mechanisms such as the sharing of nutrients and early warning for defense purposes.
- Photosynthesis delivers the fuel that propels the entire system's functions, providing the chemical energy that enables processes such as respiration. While photosynthesis has traditionally been considered as a function of leaves, trees have other photosynthetic capabilities. Many species of forest trees photosynthesize through their bark. This form of photosynthesis is generally referred to as woody tissue photosynthesis. The active chloroplast population is located mostly in the cortex (corticular photosynthesis), just under the bark, as well as in ray cells and even in pith. As the tree builds up very high internal concentrations of respired CO_2, woody tissue photosynthesis allows the tree to *re-fix* that internal CO_2 as part of its self-management strategies.
- Tree architecture is reiterative – based on the species-specific arrangement of fundamental parts such as branches, and thus part of the individual tree's management strategies in relation to its evolving environment. Individual trees, growing within a forest ecosystem, will adapt their dynamic architecture, over time, to prevailing environmental conditions and inter-spatial relationships that affect access to light, crown space or gravitational demands on structure.
- As forest trees increase in height and as competition for crown space and light increases, their lowest branches may receive less and less light and thus, their ability to maintain their own needs and to make contributions to the tree's overall upkeep may be greatly reduced. When this happens, a tree may initiate branch shedding by forming a basal protection zone, literally divorcing the branch from resources supplied by the tree, ensuring its demise. Before the dying branch is shed, all usable resources that have been stored in the branch will be reclaimed by the tree.
- Trees grow with their root crown – their buttress roots – above the soil line, as do all parts of the tree that are covered with bark. Bark, in its entirety, requires air and light in order to function properly and should never be subjected to the constant moisture of soil or mulch.

- An individual forest tree's root system begins at the trunk – root transition zone, referred to as the root crown or buttress area, with the development of at least one first order root in each of the four cardinal directions, forming the root system's perennial framework, which is primarily responsible for structural support, transport and storage. These large woody first order roots extend and branch out horizontally into the surrounding soil – the zone of rapid taper, decreasing their diameter as they grow into the fine non-woody part of the root system which, along with their symbiotic mycorrhizal fungi, are responsible for the uptake of water and nutrients. Growth is opportunistic. Root exploration of the underground landscape depends on the presence of oxygen, water, nutrient availability and a soil temperature that accommodates growth. Almost all of this extension activity occurs within the top meter/3 feet of soil with fibrous roots in the uppermost portion of this range and even into the decomposing forest floor litter, extending a minimum of 2.5–3 times the width of the crown, depending species and environment.
- Forest soil is always covered by related flora, leaf litter or other organic forest detritus – a constant recycling and renewing of organic resources. A healthy and diverse soil ecosystem – the teaming life contained within forest soils – is the foundation for forest health and forest diversity.

"For all things come from earth, and all things end by becoming earth."
Xenophanes (580 B.C.)

So, here are all these trees, growing in a forest community somewhere in the world and along comes *the hand of man*!!!

First, we built roads inside the forest creating new edges, introducing sunlight to the forest floor where it never was before, disturbing all the established networks. Then we cleared the land and evolved through cultivating it into urbanizing it. Now that more and more of us are living in cities, we've decided we want to put these forest trees back into our urban landscapes, not the way they were before, not the way they have evolved to exist, but according to our needs and our designs. While a tree's vigor is genetic, its vitality is environmental – how successfully can that forest tree grow and thrive, given the existing environmental urban conditions instigated by our needs and designs?

The fact that our urban trees, given their genetics and evolution, have found ways to deal with everything our cities and the new, very dynamic associate – us – inflict on them, one could, without any hesitation, give them an award for their resilience. They not only embody the very spirit of this word, they endure.

But, the reality of enduring is troublesome as endurance based on such impossibly contrary conditions is an act of stressed tolerance. While our urban trees struggle to tolerate the very stressful conditions in which they are asked to exist, we are not enabling their remarkable capacity to do just that. If we examine the *what is a forest* list, bullet by bullet, we can see why:

- While our urban trees are being planted into an ecosystem, it's an urban ecosystem and the ecosystem of their origin and the community that these trees evolved with and adapted to is nowhere to be found. The tree that evolved in a complex community is now asked to exist as an isolated entity, divorced from its communal associates – its family, its traditional connective tissue.

- The urban tree population is usually determined by a Landscape Architect or a Designer and the tree that's planted is not as a seed, a seedling or a sapling but as a juvenile with a DBH (diameter at breast height = 4.5 ft/1.3 m above planting grade) of 2–4 in/5–10 cm or greater, if the *design* demands a mature specimen. This is instant – there's no day, week or year. There's especially no evolution of the soil that such a tree is planted into. In fact, it is more than normal for the soil to have absolutely no relationship to the tree that's being planted, if one can even call it soil.
- As a design element, urban tree planting is only part of a much larger infrastructure construct. In most cases, these trees are planted as an afterthought, rather than planned for and accommodated. Since people are determining where trees are being planted and which species we will plant, the evolution that determined what trees are growing as associates in which forested area is no longer part of the program and, all too often, there is no direct relationship and therefore, no harmony between tree and place/ environment.
- Calling the WHAT we plant our urban trees into *soil*, one of life's fundamental building blocks, is at best, a stretch of the imagination. As previously noted, a forest soil profile with its organic layer, as well as A, B and C horizons has taken millennia to evolve. That profile has been completely abandoned in our management of the urban ecosystem. Curbside trees are planted into some sort of specified soil mixture which has nothing to do with a soil profile and which may or may not match the tree's preferred pH – the measure of soil acidity or alkalinity, determined initially by the parent material. In addition, while the soil added to the site probably has an organic component, hopefully in the form of organic compost, it is forgotten or ignored that the organic component is a renewable and in order for everything to function, the organic layer needs to be renewed regularly, whether maintained as a green ground cover, retaining healthy fallen leaves and other plant detritus or by adding composted wood chip mulch to all unsealed soils, installed 3–4 in/7.5–10 cm deep and 6–12 in/ 15–30 cm away from the tree trunk and exposed buttress roots.

  - Our open landscapes, such as Parks, are populated by trees we plant out in the lawn, ignoring the competition turf inflicts on trees. Lawn grass is a heavy feeder – it deprives trees' roots of essential moisture and nutrients, thus impacting and reducing tree root development. Additionally, most of our trees prefer a pH in the acid to neutral range. Grass wants an alkaline environment. Lawns and their upkeep reinforce this pH disparity, increasing the inability of trees' roots to access locked up nutrients.
  - When we plant trees along our streets in those curbside cutouts, affectionately known as *tree pits*, we have found a way to add additional insult to injury. The need to have safe, walkable surfaces to accommodate pedestrian access has resulted in the excessive use of impervious surfaces throughout our built communities. Not only do these surfaces seal the soil, excluding gas exchange as well as water infiltration and filtration, they have created the Urban Heat Island phenomenon resulting in re-radiated heat issues: higher local temperatures "which consequently affects building energy use, human

comfort and health, ozone production, and pollutant emissions." (Nowak, D.J., Greenfield, E.J. July 2012). https://www.nrs.fs.fed.us/pubs/40876

In addition, the engineering load-bearing specification for such installations requires that the soil substrate be compacted to the applicable standard proctor which eliminates the macro and micro pores that hold air and water and facilitate a living soil organism as well as tree root system functions. The resulting substrate does not support living systems which directly impacts on a tree's capacities and its ability to satisfy its essential requirements.

- Nursery production techniques for field grown trees often result in soil being mounded up against the trunk, which results in increased depth of roots, reducing the percentage of roots that accompany the harvested root ball. (Fig. 3.2). If the nursery is not required to remove this surplus soil, down to the root crown, before digging the tree, the result is that the tree is too deep in the harvested root ball.

Fig. 3.2 Soil mounding. Credit NC State U

When a field grown tree is dug, at least 95% of its root system remains at the nursery and that's only if that trunk – root transition zone – the root crown – is visible. The deeper the tree is below grade, the lower the percentage of roots a tree retains when it is dug, the greater the trauma, the longer the re-establishment period and the increased possibility of the development of girdling roots or root decay. In addition, if the soil that is mounded up around the trunk is not removed, this traps moisture where it doesn't belong and can lead to stem rot as trees were never meant to have soil touching any part of their anatomy that is covered with bark.

- Although a limited amount of research has been conducted on how woody tissue photosynthesis contributes to a tree's well-being, we do know that painting the trunks of trees and / or the implementation of other mechanisms which exclude light and air from stems may result in greatly limiting this functional aspect of self-management, impacting drought-stress as well as carbon availability.
- The Urban Forest also consists of many different species but selection is often primarily driven by the ability to tolerate urban conditions and does not include consideration for the environmental connections within an evolving forest ecosystem. While we sometimes successfully give trees ample space above ground, that is not the norm and we have completely disengaged from the communal landscape underground that supports the principal needs of the system and all its components. In doing so, we have not availed ourselves of trees' self-management strategies for sharing nutrients and cooperating defensively.
- Human population demands and the speed with which we develop our urban environment have established a dynamic that is quite contrary to *tree time* – the time trees require to initiate architectural adaptation. Given just how precious trees are and how important they are to our survival, we owe it to this complex organism which we only partly understand, to be observers of each unique individual we are asked to manage and maintain. We need to step back, take a breath and analyze that unique individual tree the same way we expect a Doctor to analyze what ails us. Trees have to have the time they need to adapt their architecture to the site into which they have been asked to exist instead of the instant re-arrangement we feel we need to administer. This is not to suggest that a potential for failure be ignored. What is offered, instead, is that those of us who manage these forest trees be ever mindful of how they manage themselves. That knowledge should be of primary consideration for what we do and how we do it; that we don't try to turn an oak tree into a maple but rather that species-specific plus unique individual architecture drive a health-based structure as well as its protection and preservation.

If we represent knowledge as a tree,
we know that things that are divided are yet connected.
We know that to observe the divisions and ignore the connections
is to destroy the tree.
– Wendell Berry

- When we prune, the removal of living branches has bypassed the tree's branch shedding management scheme. Because of that, the resources contained in those living branches have not been reclaimed by the tree and are therefore a loss to the tree. What is often not considered is whether or not the tree is healthy enough and has enough vitality to absorb this loss of resource. In addition, a tree's dormant buds are its *old age pension*. When trees are senescing and twigs are dying, they will often release these buds. The resulting sprouts will leaf out and provide photosynthetic resources to slow the senescing process. When we prune trees inappropriately – flush cuts or topping – the tree will release these dormant buds, forcing the use of reserves that cannot be replaced. We again come back to the need to tap into the tree's own management strategies and evaluate the tree before initiating maintenance actions which, once taken, cannot be reversed, the results of which may have negative gain for the tree and thus for us.

- Urban trees growing in a classic curbside cutout rarely, if ever, have an adequate accessible soil volume to support root system development which, in turn, can support expected mature size. Forest tree root systems can grow from four to seven times greater than the crown's diameter, forming grafted networks with surrounding trees of the same or compatible species. Compare that to the curbside coffin and its impoverished, restricted underground landscape traditionally offered to our isolated urban trees, combined with profile-less soils which do not have a maintained organic layer or horizons and which have been compacted to an engineering proctor that has eliminated most if not all the macro and micro pores which are essential for root growth.

All of this results in relentless abiotic stress. Just as we cannot remain healthy and viable if stress is an unrelenting daily occurrence, so too with trees. Our present strategies are all about tolerances. The extent of the above list shows just how far removed we are from what our urban trees have evolved with. Rather than the idea that we should ask our trees to be ever tolerant of the impoverished conditions we have planted them into, why not deliver preferences rather than tolerances? We are inserting these forest trees into our built environment – the urban ecosystem. They have not evolved genetically for this quantum leap. For such a leap to have any chance of success, the critical fact that you can take a tree out of the forest but you cannot take the forest out of the tree must be in the forefront of our management strategies.

… In a wood,
the actual visual "frontier" of any one tree is impossible to distinguish
- at least in summer.
We feel or think we feel nearest to a tree's "essence" (or that of its species)
when it chances to stand like us, in isolation,
but evolution did not intend trees to grow singly.
Far more than ourselves, they are social creatures,
and no more natural as isolated specimens than man as a marooned sailor or hermit.
Their society, in turn, supports other societies of plants, insects, birds, mammals,
micro organisms: all of which may choose to isolate and section off
but which remain no less the ideal entity or the whole experience, of the wood
– and indeed, are still so seen by most of primitive mankind….
The true wood, the true place of any kind is the sum of all its phenomena.
John Fowles, *The Tree*

The art and science of growing a tree into maturity, be it in the urban or community forest, requires a great deal of thought and wisdom. Of course, the devil, as always, is in the minute details of each and every step in the process. For our purposes, the following categories, addressing the WHERE, WHAT, HOW and WHY we will plant, establish and maintain a tree, offer a general overview of what must be thoroughly detailed:

- WHERE – evaluating and preparing the site and its soil component which, when accomplished, will support healthy tree growth into maturity;

- WHAT – selecting a tree that is true to species architecture and freed from the nursery practices which can undermine potential longevity;
- HOW – planting the selected tree into the prepared site, using only Best Management Practices (BMP) procedures, with its long term occupancy in mind and establishing the tree in its new home with an ample supply of scheduled irrigation as well as a maintained organic layer;
- WHY this is all so critical harkens back to those original connections and our extraordinary dependence on trees. Aside from the intangibles, documented research has heralded an astounding array of quantifiable Ecosystem services benefits urban trees, specifically our large, mature canopy trees, afford their human associates:

  - urban trees reduce
    - flooding and stormwater runoff,
    - energy consumption and associated emissions,
    - and the urban heat island effect;
  - they trap, absorb and filter air pollutants and particulate contaminants;
  - they improve public health;
  - and they increase
    - biodiversity – habitat for pollinators, birds and other urban wildlife,
    - carbon storage and sequestration through tree and soil preservation
    - and water filtration, infiltration, the recharging of groundwater and the resulting status of both surface and groundwater.

The non-product value of our trees has always been elusive and never given consideration until the advent of valuation methodologies such as i-Tree. Now, the Ecosystem Services benefits our trees afford the communities into which they have been planted are readily quantified via the varied i-Tree tools that are offered as freeware through the USDA Forest Service: https://www.itreetools.org

If one looks at the extent and breadth of all these criteria, it quickly becomes apparent that there is much to know and much to do. Valuation methodologies can help sell Urban Forestry to decision-makers whose largesse can then improve funding for management operations. In reality however, there will never be enough funding or staff to see to all the essentials of growing our trees into maturity, especially our large canopy trees.

Those of us who have heralded the remarkable benefits maturing trees afford their human associates have often hoped for the inclusion of the intangibles – the social, spiritual and cultural values that would complete the picture. Since these elements are subjective for the most part, quantification of such benefits is unlikely, but that's not the end of the story. The extraordinary number of participatory endeavors in both urban and community forests around the globe speaks for itself. The effort to care for trees and the landscapes they populate, initiated from a vast array of cultural perspectives and social morays, is testament to trees' intangible value and bears witness to the tree →← people connection.

With all things and in all things, we are relatives -
Sioux proverb

In approaching the sustainable management of this vital resource, this fact combined with the realization that trees belong to all of us – trees belong to people and people belong to their trees – opens the door to an additional essential element to HOW. By its very nature, Urban Forestry is inclusive, not exclusive. Growing the Urban Forest will not just take professionals but an informed all of us. Join me then, as I explore the world of participatory possibilities.

References

Nowak DJ, Greenfield EJ (2012) Trees and impervious cover in the United States. Landsc Urban Plan 107:21–30

Shigo AL (1996) How trees survive. TCI, VII, 2, (1)

Chapter 4
Growing the Urban Forest Today for the City of Tomorrow: Introducing Urban and Community Trees and Their Needs to Their Human Associates

How the philosophical premise underlying urban forestry can restore and revitalize all the residents of an urban area and beyond, be they trees or humans. What are present day mechanisms for creating and supporting the connections? An Urban Forester/Consulting Arborist's analysis of what our urban trees need and what people who care can do

Keywords Advocacy · Building WITH trees · Informed actions · Participatory management · Planning · Protection and preservation · Site evaluation · Social cohesion · Tree inventory · Tree planting · Tree selection

Fig. 4.1 *Platanus x hybrida* at Avenue Henri-Dunant Geneva. 2017. Naomi Zürcher

Supplementary Information The online version contains supplementary material available at [https://doi.org/10.1007/978-3-030-94534-3_4].

N. Zürcher, *Connecting Trees with People*, Future City 16,
https://doi.org/10.1007/978-3-030-94534-3_4

The tree in front of my home is a word,
the trees on my street are a sentence,
the trees in my neighborhood are a paragraph
and all the trees in the community are a story.
That story tells us about our relationship to nature, past and present.
The future of this story lies in the hands of all residents…
~Greg McPherson,
Supervisory Research Forester,
Pacific Southwest Research Station,
USDA Forest Service, Davis, CA.

The connections between trees and people and how those connections have been realized over millennia, as described in Chap. 1, have established just how interconnected we are to trees and thus to the land in so many very primal ways. The human story is anchored in an agrarian dominated lifestyle and although we have stepped out of that story, I believe it still remains in our genes, our memory and in our psyche.

Does the fact that these connections might be dormant but reachable constitute the structure for a non-linguistic way of communicating, such as the "Language of the Birds" (https://en.wikipedia.org/wiki/Language_of_the_birds)? Globally, music has been seen as THE universal language – regardless the language of origin, music speaks to the elemental emotions in each of us, beyond the differences in the way we speak and the way in which we culturally make music. The land is similar. Although stewardship strategies differ from culture to culture, we recognize in each other the bond that caring for the land, the earth, creates. This universal language in stewarding the earth can break down barriers between communities and can create social cohesion, connect citizen with immigrant, establish a basis for common ground. Trees are a pathway into the earth, a path worth taking.

What do we wish?
To be whole. To be complete.
Wildness reminds us what it means to be human,
What we are connected to rather than separate from…
This is about putting ourselves in accordance with nature,
Of consecrating the lands by remembering our relationships to them…
Terry Tempest Williams

The analyses of forest tree and urban ecosystem have laid the groundwork for Urban Forest inclusiveness as it is applied to participatory management. To build on this foundation, we need to explore and understand these two words – participatory and management – both conceptually and functionally, within an Urban and Community Forestry framework – and then connect and merge them into a seamless synergistic approach.

Participation has been defined by the Oxford online dictionary (https://www.lexico.com/definition/participation n.d.) as the *act of taking part in something*. In many reviews and analyses of the public's participation, the role assigned to the public has been primarily advisory with no guarantee that the advice will have an effect on outcomes. In April of 2002, the Ministerial Conference on the Protection of Forests in Europe published a synopsis of findings and recommendations by the FAO/ECE/ILO (Food & Agriculture Organization/United Nations Economic

Commission on Europe/International Labor Office) Joint Committee Team of Specialists on Participation in Forestry, entitled "Public Participation in Forestry in Europe and North America" https://foresteurope.org/wp-content/uploads/2016/08/public-participation-in-forestry.pdf.

The *What is…* definition for the subject at hand, according to The International Labor Office, Geneva, was stated as "Public participation is a voluntary process whereby people, individually or through organized groups, express opinions and articulate interests, and have the potential to influence decisions or the outcomes of the matter at hand." On the contrary, true participation is equality, sharing power, having decision-making responsibilities. Although this report was published in 2002, more than a decade after the USDA Forest Service implemented its nationwide Urban & Community Forestry program, there is no mention in this report about this North American citizen participation program nor its emphasis on hands-on participation.

In addition, we are facing a monumental climate crisis, with Climate Change impacts escalating faster than previously forecast. We are presently confronted with the accumulation of *us* and what that has wrought. We in the west have spent the last 200+ years overindulging in the Earth's bounty, living as if the barrel had no bottom. To start, we lived off the fat of the land. We have worked our way through the fat into the bone – the skeleton. Our excesses are undermining our collective structure and what is needed to maintain and sustain structural integrity. While there are those who are still obfuscating or are in denial, the calamity is accelerating and expanding and the precariousness of our situation is palpable.

Our own irrelevant preoccupations can be observed in how we care for our trees. Although trees support and enhance a vast array of diverse life including us, we, for the most part, take them and their place in our world for granted until confronted with their absence. Sometimes, it's the loss of an individual tree – the one we grew up with, climbed in, hid behind. Maybe it was due to old age or disease, a devastating storm or the developer's backhoe. Sometimes it's the mass of the loss – an infestation, a natural disaster or a conflagration of massive proportions – which irrevocably alters the entire landscape, turning verdant streets or hillsides into moonscapes.

There are many people who want to do something that will make a difference, not just with words and opinions but with informed actions. Given that a great deal of the negatives impacting our environment are generated in our urban areas, we need to create opportunities in our urban areas where those who would can and those who can know how and why.

Since our focus will be participation in management in its broadest sense, we need to first look at the precursors for a participatory structure in the art and science of growing an urban tree. We then have to define the trees' needs-based management protocols and practices from planning, design and tree selection to planting, establishment and maintenance with a protection, preservation and advocacy focus, to see the WHERE, WHAT, HOW and WHY of participation possibilities. If we incorporate *inclusiveness*, the philosophy inherent in Urban Forestry, into the formulation of our scope of possibilities, we will see the larger benefit in supporting citizen participation.

First and foremost, we need to establish that the Urban Forest we are discussing should not begin somewhere *out there* – a place somewhere distant where we can go to seek respite from our urbanity. This statement is not intended to diminish the incredible value of forests or other aspects of the natural environment or suggest that we should not enjoy these spaces passively and with respect for all the residents. The protection and preservation of naturally-occurring ecosystems is critical to the existence of Planet Earth and all her inhabitants. But, many of us do not have the capability or the financial means to access the magical green spaces and places that are *out there*. Since the Urban Forest belongs to everyone, it should be for everyone, beginning with the place we call home. Our streets should be places you go to, not places you go through on the way to someplace else to find that respite. The Urban Forest needs to begin at every person's doorstep, in every person's neighborhood, as a vibrant part of our community. And, just as we make decisions about our home's interior, we might think of what's just beyond the door as our home's exterior, but our home none-the-less, for isn't that what a neighborhood is?

Broad-streeted Richmond...
The trees in the town are old trees
used to living with people,
Family-trees that remember
your grandfather's name.
Stephen Vincent Benet
excerpted from John Brown's Body

Urban Forestry has become an essential ingredient for a healthy sustainable urban area in part because it grew out of a grassroots effort – bottom up, not top down. It is inclusive, not exclusive, and it's growth was thanks to the broad brush approach to planning and implementing such a forest. It is quintessentially URBAN if you think of urban as people – people and our stuff are its definer and it is people – stakeholders – from all different perspectives that have made Urban Forestry live wherever its seeds have been planted.

Who are the stakeholders, the *owners* of the Urban Forest? They – we – are all of us, regardless of age, religion, nationality, gender, sexual preference or economic situation. We are students, workers, professionals (including tree experts), scientists, teachers, retirees. Our common ground is that we care and want to donate our time, energy and, in some cases, our expertise to the well-being of the urban ecosystem. And our trees? They don't care who stewards them, only that you care to do so.

And, since the Urban Forest belongs to all of us, participation which only recognizes a *potential* to affect decisions does not inspire. While opinions and interests are important and valuable, the ability to actively steward the resource – trees, the land and each other – is a critical ingredient in the process of taking ownership. What should always precede participation in such an endeavor is a basic understanding of the what – what is a tree, what are its need and what can we do to best accommodate those needs – along with related applied learning opportunities that take the learning and put it into the ground. The successful stewarding of such a sophisticated organism requires basic knowledge and skills, some of which have been offered in preceding pages.

There are numerous ways in which this basic knowledge can be shared. Initially, information can be made available through web portals but these important communications are best offered in person through educational workshops where professionals and citizens can meet, information can be shared and an open dialogue initiated. These workshops can facilitate connections between official managers and prospective citizen stewards and establish the basis for a working relationship. They can also initiate connections between prospective citizen stewards from varied communities, cultural backgrounds and social groups.

Why is this education so critical? The community needs to have a voice as well as actionable participation in all aspects of planning and management. That voice needs to resonate with management officials so that whatever initiatives the community offers are seen to be cogent with the weight of knowledge behind it. It is critical that when our initiatives realize the goals we have set, they provide benefits for our trees as well as ourselves.

The outcomes of such initiatives are expansive. A community can develop self-reliance through collective experiential projects. We learn best by doing and when our doing is communal, we grow connections that are very similar to trees' root systems. We form a web – an inter-connected network – which enables a shared quality of life and fosters well-being.

Assessing the Resource – Tree Inventories Plus i-Tree

The planning for and management of any tangible resource begins with an assessment of what already exists. This is no less true for the urban tree resource. "Regardless of the questions that Policy-makers, Planners or Urban Foresters might pose from their varied perspectives, answers can only be provided if the resource has been accurately quantified. Quantifying begins with an urban tree Inventory – from deciding on objectives and type of Inventory to using the resulting data for management: policy-making, land use planning, design, maintenance and valuation of ecosystem services…" (Zürcher 2017)

Traditionally, tree inventories focused on data collection to facilitate maintenance priorities and scheduling as well as species diversity and available planting locations. The current emphasis on Ecosystem Services benefits and the ability to quantify those benefits through i-Tree Eco tree assessments has expanded the data to be collected and its uses in planning and managing the resource.

The following Table 4.1 Inventory Data to be Collected (Zürcher 2017) provides an overview of the data that should be collected during an urban tree inventory – basic data, additional essential data as well as subsequent management data – and includes those parameters which are part of an i-Tree Eco assessment.

The Table also includes a 4th column calling out those elements of an assessment which observant, informed citizens could readily and reliably collect, especially when this is done in partnership with the municipality. Urban tree inventories are the second most popular participatory action in Urban and Community Forestry. Participation in a tree inventory is an excellent way to encourage tree observation skills.

Table 4.1 Inventory data to be collected. 2017, 2020. Naomi Zürcher

	A	B	C	D	E
2	**Inventory Data to be Collected**	**Basic Data**	**Additional Essential Data**	**Subsequent Management Data**	**Citizen Participation**
3	**Location**				
4	Physical - the address being surveyed	✔			★
5	GPS coordinates for GIS mapping	✔			★
6	**Spatial Resource: Site Characteristics Condition**				
7	Land use or site class [⌘]	✔			•
8	Planting area type		✔		
9	• tree lawn / grass strip		✔		
10	• curbside cutout		✔		
11	• street mall / median strip		✔		
12	• parking island		✔		
13	• urban park		✔		
14	• open field		✔		
15	• urban woodland		✔		
16	Planting area dimension		✔		★
17	Planting area treatment		✔		★
18	• open accessible soil volume		✔		
19	• permeable or impermeable pavement / surface		✔		
20	• ground cover		✔		
21	• open landscape		✔		
22	Paved walkway condition		✔		★
23	Amount of street traffic		✔		★
24	Presence of overhead wires		✔		★
25	Presence of underground utilities		✔		
26	Adjacent use				★
27	• buildings including type and size e.g. one-family, multi-residence, business		✔		
28	• parking lot		✔		
29	• public or private greenspace		✔		
30	• vacant land		✔		
31	**Tree Resource: Current**				
32	General Tree Metrics				
33	• Genus and species [⌘] + cultivar or variety, if known	✔			✚
34	• DBH (diameter at breast height, specific or class) [⌘]	✔			•
35	• total tree height [⌘]		✔		★
36	• total live tree height [⌘]		✔		✚
37	• height to crown base [⌘]		✔		★
38	• crown width [⌘]		✔		★
39	• percent crown missing [⌘]		✔		✚
40	• crown light exposure [⌘]		✔		✚
41	• proximity to buildings within 60'/18.3m [⌘]: direction N>S, E>W; distance		✔		★
42	Condition				
43	• incremental annual growth		✔		
44	• percentage of dieback [⌘], foliage transparency, crown density, live crown ratio	✔			
45	• risk assessment and rating	✔			
46	• mechanical injury		✔		
47	• pathogens	✔			
48	• abiotic impacts		✔		
49	• condition notes e.g. specific observations affecting maintenance		✔		
50	Conflicts with Infrastructure				
51	• pavement [⌘]	✔			★
52	• buildings	✔			★
53	• overhead wires [⌘]	✔			★
54	• street / roadway signage	✔			★
55	• street / roadway lighting	✔			★
56	Actions recommended				
57	• maintenance, e.g. pruning	✔			
58	• removal	✔			
59	• risk mitigation	✔			
60	• site mitigation, e.g. increase tree pit size, bevel pavers	✔			
61	**Tree Resource: Subsequent**				
62	New Planting Data				
63	• GPS / GIS coordinates			✔	
64	• Genus, species, cultivar / variety			✔	
65	• Provenance, if known			✔	
66	• date planted			✔	
67	• caliper / height at planting			✔	
68	• production type e.g. field grown, container grown			✔	
69	• package type e.g. B&B, bare root, container specifics			✔	
70	Public Construction Inventory Data (to be submitted at Project's conclusion)				
71	• GPS / GIS coordinates			✔	
72	• Genus, species, cultivar / variety if known			✔	
73	• planned intervention e.g. clearance pruning, root pruning, transplanting, irrigation, removal, including dates of occurrence			✔	
74	• documented - dated construction impacts / mitigation e.g. soil compaction			✔	
262					
263	[⌘] - i-Tree assessment parameter				
264	[★] - unassisted participation				
265	[✚] - assisted participation				

Planning and Design

The materials of city planning are:
sky, space, trees, steel, and cement;
in that order and that hierarchy.
~ LeCorbusier,
Swiss architect and city planner

As our forest-cum-urban tree is a very complex organism with many self management strategies in place, the concepts required to manage this resource within the setting of urban spatial development, as generally described in Chap. 3, have many parts and pieces. The detailed procedures for planning, designing and preparing a tree-worthy urban site are numerous and the devil is always in the details, as enumerated by the following list:

- what type of planting area is it – open landscape, tree lawn, curbside cutout?
- what are the proposed dimensions of the open tree area?
- are there other trees already existing vis-à-vis the planned planting site; what species of trees already exist; what is their condition; what soil pH do they prefer; is their mycorrhizae species endo or ecto?
- are there grading issues – too high or too low?
- are there overhead wires or underground utilities?
- what is the proximity to grey infrastructure – curb, pavement, buildings; existing or planned – how tall are the buildings, what is the proposed planting site's orientation to the buildings as well as to the street?
- does the pavement already exist – what type of paving material; what is the condition of the material; does the new planting include sidewalk replacement?
- will there be re-reflected heat load issues or is the site particularly windy which can contribute to droughty conditions, excessive transpiration or reduced photosynthesis activity?
- what is the daylight availability throughout the growing season?
- how much open, accessible soil volume will the tree have – THE most limiting factor to an individual urban tree's ability to thrive and thus be the amenity provider and ecosystem services benefactor we are counting on;
- and, what about the soil – to determine the right tree for the site and if the existing soil can be retained, conduct a thorough soil assessment for pH, chemical contaminants, drainage and bulk density issues;
- what is the total accessible soil volume, unsealed and sealed; are there possibilities to expand rooting space; for example, joining individual curbside cutouts into one continuous planting strip supports increased root development, facilitates a communal rhizosphere and could accommodate root paths or soil trenches to increase the volume of un-compacted, accessible soil;
- if the site is a curbside cutout and the area across from the pavement is open landscape, structural soil strategies such as CU® or Amsterdam soil which, when compacted, retain their structure, could be used to convey the roots under the pavement to the larger, open soil volume and, at the same time, reduce tree – sidewalk conflicts.

Many of the site preparation procedures, such as installing root paths to increase root system access to a neighboring tree's root system and collectively to a greater volume of soil, involve highly technical tasks related to designing the construction or reconstruction of a walkable surface installation.

But, the planning for and designing of the community's Urban and Community Forest should involve the community wherever and whenever *it's MY community* has potential. In addition to the municipality's Urban Forester, a representative from the official entity responsible for planning oversight should reach out to community leaders to initiate community planning meetings/ listening sessions that consider community history and current trends in order to:

- establish procedures for participation in the particular issue to be discussed;
- learn about cultural issues within the community which might affect receptiveness to trees being planted and retained, e.g. residents of Chinese heritage may need to have Feng Shui principles applied to sites chosen for tree planting and which trees should or should not be selected;
- share the planning and design strategy for the prospective tree installation;
- discuss proposed development in relation to what currently exists; what has worked well from the community's perspective in terms of tree longevity and minimized conflicts and what has not;
- review how new development might impact the community's existing tree and green space resources and potential strategies for the mitigation of such impact;
- listen to the community's thoughts and priorities: suggestions, concerns, requests;
- clarify misconceptions regarding trees and potential conflicts, often caused by inadequate site preparation and resource maintenance;
- define community's objectives and goals and its willingness to participate in their realization.

Popular modalities such as questionnaires, interviews or surveys provide some superficial information but these cannot replace the group dynamic and what that elicits. Facilitating a community's hands-on participation in the creation of an environmentally sustainable shared vision facilitates place-making while enhancing environmental well-being as well as the health of the community's residents. The collective outcome of planning and design listening sessions must then find its way into a Strategic Plan, capturing the resulting vision through goals, objectives and policies that have been determined and appended with the updatables – Urban Forest Inventory, Standards of Practice, etc. This entire process of visioning a city is exemplified by Winnipeg Canada's Speak Up Winnipeg project, described in Chap. 5.

Tree Selection

As with planning and design, there is a great deal of expertise required in the selection of a tree capable of taking up residence and growing into maturity at a designated urban location. Once the spatial resource has been evaluated and prepared, a site-specific tree list can be developed. While ornamental attributes are always important to designers as well as the public, the health of the tree material and its applicability to the site into which it will be planted and asked to grow must come first. And, while communities can contribute their wishes during the planning and design phase, the actual decision of what should be planted where should be left to those with the qualified experience and expertise. Preferences, tolerances and conditions must be part of developing the site-specific list, be it for an Urban or Community Forest planting as follows:

- cold hardiness and heat preferences – even trees with a broad hardiness range should be selected based on the specific local provenance as genetic variables can affect survivability; if the tree is grafted, the grafted root stock must be cold hardy for the site being planted;
- light preferences – the amount of sun the selected tree will require in order to grow into maturity – full sun = direct sunlight for a minimum of 6 hours per day; partial shade = 3–6 hours of direct sunlight per day; shade = less than 3 hours of direct sunlight per day;
- soil preferences – the soil conditions at the prepared site must match the soil preferences for the tree being selected for moisture/drainage and soil characteristics, e.g. structure, texture, pH;
- varied tolerances – site conditions such as drought, air pollution as well as the use of de-icing salts must be considered to ensure that the selected tree can tolerate the presence of such conditions;
- allergies – if the tree will be planted in a community with a high incidence of asthma, select trees that do not exacerbate the condition but have a high rating for air pollution mitigation, a primary causal agent of asthma;
- biodiversity – if the community receiving the tree has an interest in increasing the biodiversity within their community, include trees that offer habitat or forage for birds and pollinators;
- restrictions – rooting space, above ground utilities - if the site has restrictions which cannot be mitigated or eliminated, avoid trees that have known rooting space issues such as *Tilia* or *Fagus* species or height/crown form which will compete with over head cables;
- planting schedule - avoid selecting trees for Fall planting that should not be dug in the Fall, such as conifers or *Liquidambar*, as noted in *Noteworthy Requirements for conifer and hardwood selection*.pdf. (Zürcher 2007)

Certain nursery production practices can greatly influence the percentage of roots which will accompany the harvested tree and therefore, if not understood and addressed, can lead to decreased longevity and increased maintenance costs down the road. Understanding how these practices impact trees over the long term will help facilitate the tagging and harvesting of quality material:

- soil is often mounded up against the trunk of field grown material, which impedes examining the root-trunk transition area and, more importantly, results in increased depth of roots, reducing the percentage of roots that accompany the field dug root ball. The nursery must be required to remove this surplus soil, down to the root crown, before digging, to ensure that the harvested ball has roots where they should be and not at the bottom of the ball (Chap. 3, Fig. 3.2);

- heading back trees to stimulate branching can result in undesirable groupings of branches originating from the same point in the tree. Only select trees that have branching true to species' characteristic architecture;
- use field-grown or bare root material, whenever possible - container grown production techniques have not yet resolved root circling and girdling issues adequately enough to ensure quality standards can be met.

Tagging. Trees must be selected and tagged by a tree expert or equivalent at a growing nursery located within the cold and heat hardiness zones of the location to be planted. A full examination of any tree being considered, as noted in *Tree Shopping Guide – selecting quality in the nursery*.pdf (Mulcahy and Zürcher 2007) should include:

- the tree's crown, consisting of trunk, branches, buds, characteristic of the species and of exceptional quality;
- trees with co-dominant stems and/or included bark should never be tagged;
- the root-trunk transition area should be examined for possible girdling roots, damage, or injury;
- be sure to locate first order roots with a surveyor's pin - good root structure, critical to tree health, should consist of at least one permanent first order root, in each of the four cardinal directions, located within the top 3"/7.5 cm of soil;
- a grafted tree's graft union must be visible. If planted too deep, the grafted portion can begin to root, weakening the graft structure as well as disease resistance;
- mark selected trees on their north side for transplanting orientation to reduce the incidence of sunscald.

Where participation in the tree selection phase is possible and should be encouraged is determining which trees from the prepared list the community would feel drawn to, understanding not only the aesthetic or cultural contribution but the Ecosystem Services benefits the selected tree can afford the community if the tree is grown into maturity.

Tree selection for the restoration of a Community Forest is not nearly as complicated, as one is normally dealing with seedlings or saplings that are native to the location. It is still important that the material selected is representative of species-specific architecture and free of visible defects.

Tree Planting and Establishment

A tree growing out of the ground
is as wonderful today as it ever was.
It does not need to adopt new and startling methods.
– Robert Henri

Tree planting is probably the most common aspect of participatory Urban and Community Forestry. While tree planting is often thought of as a simple act (Lipkis, A & K. 1990. *The Simple Act of Planting a Tree*. J.P. Tarcher), if one looks at the extensive array of details which must be addressed, it is, in actuality, fairly complicated and involved. A field grown, balled and burlapped (B&B) tree, destined for the street, would rarely be smaller than 2.5"/ 6.35cm caliper. The average weight of such a tree ranges from 400 to 500 pounds/ 181 to 227 kilograms. That's a lot of tree to manage, given that it can only be lifted by the root ball. Why is that, you might ask? It has been suggested that the soil ball causes a 90% increase in weight over the same tree bare root (soilless). If a B&B tree is lifted by its trunk, gravity will cause the soil (where most of the weight is) to break up and pull away from the roots which can greatly affect the ability of the tree to grow as one can rarely put the ball back together adequately enough to re-establish the soil – root union.

Successfully planting such a field grown tree into the prepared site requires adhering to all of the following procedures:

- measure the root ball height prior to digging the planting hole, ensuring the tree's root crown will be at or slightly above planting grade and the ground underneath the ball will remain undisturbed;
- planting hole sides must be sloped so the top of the hole is a minimum of 2.5–3x the width of the root ball;
- soil should be removed in 1/3 lifts; store the removed soil lifts on geotextile fabric and do not break up soil aggregates;
- prune only damaged branches back to a lateral before lowering the tree into the planting hole;
- always lift the tree by its root ball, never by its trunk;
- center the tree plumb in the hole with the marked side oriented north – this reduces the likelihood of sunscald;
- reversing the soil removal process, backfill with retained soil to 1/3 the root ball height;
- cut the wire basket, if present, and untie and cut burlap, removing both to this backfilled height, giving roots immediate access to surrounding soil;
- rough up the sides of the planting hole as well as the root ball to reduce the soil interface but do not destroy or fracture the root ball;
- water in the backfilled soil, continue to backfill in lifts of one third, watering in each third; do not place any backfill on the top of the root ball;
- using soil, create a saucer-like water basin at the outermost edge of the root ball to encourage water to flow into the ball;
- in situations where the tree needs to be stabilized, instead of above ground staking, opt for a root ball staple – an underground, decomposing root stabilization system made of untreated wood – so that only the root ball is stabilized while not impeding movement in the above ground part of the tree, necessary for development of trunk taper;
- install composted wood chip mulch, 3–4"/7.5–10 cm deep, 6–12"/15–30 cm away from the tree trunk;
- DO NOT FERTILIZE.

An excellent resource for the soup to nuts protocols for planting urban trees is Watson, G. W, Himelick, E. B. 1997, 2013. *Principles and Practice of Planting Trees and Shrubs*. International Society of Arboriculture.

While container grown material is lighter in weight (as a result of being grown in a soilless medium), plantable year round and seemingly easier to handle, containers present their own assortment of problems and complications which must be overcome if the tree that's planted will be able to grow into maturity.

One of the supposed advantages of container grown and containerized material is that there is a much more complete root system accompanying the tree. The difficulties reside in the packaging which causes the root system to circle around itself in addition to roots that are ascending, descending and/ or kinked. These problems are usually not corrected each time the tree is stepped up into a new container size and thus, the problems are not only along the surface of the root ball but are buried deep inside the ball. In addition, the container grown tree might have originated in one nursery and then shipped to another. When that second nursery stepped up the tree, the surface root problems might have been addressed but not the ones located further inside the ball. These issues need to be recognized and corrected in order for the tree to grow and to have the necessary root stability to keep it upright and healthy. Dismantling such a root ball using techniques such as root washing should be done by someone with experience as it is not for the faint hearted.

What lends itself most to community planting initiatives is bare root material. There are certain advantages but also disadvantages. The advantages are numerous including:

- more manageable due to greatly reduced weight;
- much less expensive than purchasing and shipping a field grown, container grown or containerized tree;
- a much greater percentage of roots are harvested with the tree, greatly reducing transplant trauma;
- the all-too-often buried trunk flare syndrome is completely eliminated; in addition, the entire root zone is visible and problems such as kinked or damaged roots can be addressed and eliminated prior to planting;
- as there is no root ball, there is also no soil interface problem to overcome and the root mass accompanying the tree has immediate access to the backfilled soil and its moisture content.

Most of the disadvantages of bare root material have to do with limitations to:

- planting season – bare root trees MUST be dug and planted when dormant;
- the number of species which have proven success using this methodology; trees that are intolerant of desiccation are not good candidates for bare root;
- the available caliper size range – usually between 1.5 and 2"/3.8 and 5.0 cm; this is often too small to meet municipal curbside parameters for minimum caliper;
- timing between digging and planting – bare root trees cannot be held for more than a week once they have been dug including shipping time;

- storage techniques and practices, as the lack of soil around the roots can instigate root desiccation fairly quickly; strategies such as a hydrogel dip immediately following digging reduces the incidence of desiccation but does not eliminate the problem and so bare root trees require constant shade and moist conditions during their brief time out of the ground.

Even with the disadvantages, bare root is an excellent planting methodology for community planting projects. An excellent resource with detailed procedures, entitled "Creating the Urban Forest: The Bare Root Method", is available for download from the Urban Horticulture Institute, Cornell University, at the following website http://www.hort.cornell.edu/uhi/outreach/pdfs/bareroot.pdf along with the video "Creating an Urban Forest: The Bare Root Method", which can be viewed at https://vimeo.com/9729714.

The contribution community tree planting initiatives has made to social cohesion, social network systems and community empowerment has been well-documented in varied research projects (Fisher et al. 2015; Sommer et al. 1994a, b; Westphal 2003; Watkins 2017. https://doi.org/10.1016/j.cities.2017.11.006)

The questionnaires and surveys conducted to establish the benefits citizens realize from these initiatives are proof positive as to the physical and mental benefits such experiences initiate. What is rarely discussed, however, is if citizens' efforts have resulted in the enabling of planted trees to grow well into maturity. Were all the devilish details required to successfully grow an urban tree part of the endeavor? We all-too-often think about nature and its processes as being there to serve us, harkening back to those biblical edicts referenced in Chap. 2. Citizens' willingness to roll up their collective sleeves and to make a difference by getting involved needs to be channeled in such a way that the efforts they make DO make a difference. The guiding principal in all of these actions needs to be *will it benefit the tree?* If it benefits the tree, it will most assuredly benefit the community.

Establishment – when a field grown tree is dug, 95% of its root system remains at the nursery and that's only if the root flare is visible. Essential establishment practices can help the tree to quickly acclimatize to its new location – specifically putting on new roots to reduce transplant stress:

- post planting watering providing 20 gallons/75 liters of water per week must be part of the planting contract and each watering must be verifiable; tree irrigation bags such as Tree Gators or their equivalent have proven to be a reliable tool to provide slow, consistent, verifiable supplemental irrigation to newly planted trees;
- a composted wood chip mulch installation, as detailed above under **Tree Planting**, must be maintained to mitigate soil compaction and erosion, retain soil moisture, stabilize soil temperature and provide the organic layer necessary to support the soil microbial community;
- if the tree has been contract planted and stabilized using above ground staking material, removing such material prior to the end of the guarantee period could void the guarantee; that does not preclude checking the stake guy ropes to make certain they are not girdling the tree's trunk.

Establishment has the potential for critically important participatory efforts. Wherever contracted plantings are involved, the eyes and ears of an informed and interested community are invaluable. Are those irrigation bags being filled on a weekly basis? Are the bags not emptying because the weep holes are clogged? Is the wood chip mulch installation being maintained correctly? If stakes and guy ropes were used, are the guy ropes causing damage? This is such a pivotal time for the transplanted tree and its need to acclimatize as quickly as possible to its new home. The delivery of essential establishment practices can be assured by an attentive and proactive community.

If the tree has been planted by the community, responsibility for establishment procedures, as noted above, should be determined by the municipality and delegated accordingly. If the responsibilities have been delegated to the community, the municipality should provide any and all materials required to fulfill that obligation and the community allowed to devise an appropriate schedule and assign the tasks to ensure that it can deliver the delegated responsibilities.

Post Planting Care and Long Term Maintenance

Respect trees – trees have dignity too.
Learn about trees and their associates
so that you can help make better decisions
for their long-term, high quality survival.
Dr. Alex Shigo, father of Modern Arboriculture

We have clearly established that trees growing within a forest ecosystem have the ability to self-manage and maintain their collective well-being. That's not quite the case for the forest-cum-urban tree and so we have arrived at the not-so-sexy critical aspect of GROWING the tree that has been planted into the built environment.

There has been much written about the cost of NOT maintaining the Urban Forest. From 18–20 March 2015, the International Society of Arboriculture Science and Research Committee hosted a research summit, *The Costs of Not Maintaining Trees* (CNMT) in Tampa, Florida, U.S. The focus of this summit was to define research questions related to the economics of tree care which could establish appropriate levels required to "efficiently maintain tree health and structural integrity". (Koeser et al. 2016.) Additionally, we have a review of all the arboricultural and urban forestry literature detailed in *The Costs of Maintaining and Not Maintaining the Urban Forest: A Review of the Urban Forestry and Arboriculture Literature*. pdf (Vogt et al. 2015). This review has unearthed some interesting facts about what is considered in the cost of maintaining our urban trees.

In the section captioned "Maintenance Not Commonly Quantified in Municipal Budgets", we find entries such as watering, mulching, soil and nutrient management and protection during construction. If we look back at forest tree structure and function, as described in Chap. 3, we can already see where after care and maintenance have ignored essential basics. For example, given that healthy soil is a prerequisite for healthy trees and that one of the prerequisites for healthy soil is the organic layer without which the microbial community cannot exist, it is quite extraordinary that these critical cost-effective elements are not part of a Municipal budget. It is also quite evident that most municipalities are not only failing to incorporate essential basics into their after-care protocols, they are not attending to most of the other critical procedures, from site evaluation and preparation onward. What we do or don't do up front will determine how much we have to make up for down the road and what happens in the *make up for* category is always much more costly than if we had dealt with all the necessary protocols, procedures and best management practices to start with, as itemized throughout this Chapter. Not only do we incur greater expense, but we are also incapacitating our trees, that invaluable resource.

While there are many components of after care and maintenance which must be relegated to tree experts – risk assessments, pruning for tree health, treatment for pathogens, etc., the essential basics of watering and mulching are perfect projects for citizen participation not only because they are not being attended to with any regularity but because of just how vitally important they are to growing an urban tree into maturity. The essential basics - how to water a tree during droughty conditions, how to incorporate organic compost into the soil and how to mulch a tree correctly require minimal instructions. Community leaders and/ or associations can usually be inspired to take on such responsibilities and disseminate them throughout the community. With materials and basic instructions supplied by the municipality, these critical tasks can afford communities experiential opportunities that truly make a difference.

Preservation and Protection

Conservation is a state of harmony
between men and land.
Aldo Leopold

While all the varied components of management reviewed thus far are part of preservation and protection, there are aspects that do not fall into any of the above categories. The steady influx of people into our urban centers combined with population growth and how that growth has led to urbanization in what was once rural has expanded the urban center footprint exponentially. Urbanization instigates spatial development. Since land that is readily available for urban development is greatly limited, developers often look to green space as *vacant* land and, therefore, available for building. While a surging population demands living spaces and services, human well-being demands green spaces and places – a virtual paradox – how to accommodate spatial development demands while sustaining human needs. (Fig. 4.2)

Fig. 4.2 Paradox. Aubervilliers neighborhood. Paris France. 2015. Pablo S. Herrero

> "Managing urban areas has become one of the most important development challenges of the 21st century. Our success or failure in building sustainable cities will be a major factor in the success of the post-2015 UN development agenda," said John Wilmoth, Director of UN DESA's Population Division https://www.un.org/en/development/desa/news/population/world-urbanization-prospects-2014.html

Although Urban and Community Forestry cannot resolve all the issues which play to the paradox, we can, through protection and preservation concepts, ordinances and procedures, amplified by informed citizen advocacy, make the playing field more level and balanced. As with planning and design, protection and preservation can readily involve an informed citizenry. Collaborative initiatives known as Tree Boards (or Committees, Councils…), empowered by a local Tree Protection Ordinance or a local Executive Order, and usually incorporating Municipal personnel, tree professionals and informed citizens, can help to shape Urban Forestry Management policy and administration. Courses that provide excellent resources to learn more about this aspect of protection and preservation-focused participatory management can be found at the Tree Board University website https://tree-boardu.org

Although Tree Boards are, for the most part, an American phenomenon, the concept and its implementation make for an excellent participatory model on which to build a more sustainable urban forest ecosystem, no matter where it is located.

Local tree protection ordinances can facilitate the protection and preservation of the resource but only if the ordinance is enforced which is often not the case. No matter how much money is available for Urban and Community Forest Management, no matter the staff capacity, there will never be enough eyes and ears to protect and preserve the entirety of the Urban Forest resource. The potential for unintentional harm is great and, in defense of our urban trees, the need for informed community involvement is an imperative.

An informed and engaged community translates into excellent observers. That shared basic understanding of the *WHAT* we discussed earlier in this chapter will provide the essentials necessary for observations which can save trees. Construction equipment invades existing trees' structural root areas, tearing into the soil, fracturing roots back to the trunk - initiating root decay and, potentially, root rot disease. The vast assortment of projects ranging from public infrastructure - sidewalk and curb replacement, above and below ground utility management, local street and parkway rebuilding, school construction – to private activities - driveway installations, residence construction/reconstruction – is, more often than not, impacting publically-owned trees which happen to fall within or in proximity to the project's work footprint or staging area. The damage such work inflicts on the below ground parts of a tree is often not discovered (if ever) until it is too late. Community involvement can often short circuit negligent practices by reporting early abuse or lack of appropriate protection measures as the project is being established. An online resource showing common construction project abuses can facilitate community observations. It is critical that reported observations be acted upon promptly, before damage is irreversible. Such actions go a very long way towards making protection of our trees a reality. When Contractor's know they are being observed, they often act more responsibly.

An essential part of protection and preservation is the implementation of "Building WITH Trees" strategies, using construction techniques and applying them to protecting and preserving trees and the landscapes they populate. (The following has been described, in depth in the Chapter entitled *In Consideration of the Tree: The importance of structure and function in the realization of Ecological Design* [Zürcher, N. 2021] in the Springer Nature COST project book *Urban Services to Ecosystems: Green Infrastructure from the landscape to the Urban scale*)

- once a project's work footprint has been defined and agreed upon, a Tree and Landscape Protection Fence must be installed to define the Tree and Landscape Protection Zone;
- when construction vehicles require controlled access to protected landscapes, soil compaction and impact can be greatly reduced by installing Ground Protection Mats or a Portable Road on top of a 12 inch/30 cm layer of wood chip mulch over geotextile fabric;

- pneumatic/ hydro excavation, beginning with vacuum excavators, has been used for more than a century by the construction industry for digging in certain situations such as mining or when installing or working around utilities. Arborists recognized the tool's non-impact/ non-destructive possibilities for use in tree-related excavations when a root diagnostic nozzle has been incorporated. Its present construction-related usage includes exposing root systems to facilitate precision root pruning as well as radial trenching to mitigate soil compaction;
- trenchless technology in the form of microtunneling or augering, has been used to install pipes and conduits under roadways and the like. On projects that will impact a large number of healthy mature canopy trees, trenchless technology can greatly reduce the number of trees that are slated for removal. Microtunneling under trees' roots can eliminate the extensive damage to root systems as well as the entire landscape, thus enabling tree retention and reducing the costs required to restore the landscape;
- construction activities also damage the rootable environment – the soil. Concrete washout adds contaminants to the soil which are toxic not only to trees' roots but to the entire living soil organism.

While these strategies are highly technical and beyond the scope of the invested community, sharing the strategies which the municipality is incorporating into its project requirements is a strong indicator that the municipal government is taking proactive steps to help ensure the longevity and well-being of the resource. This, in turn, gives citizens greater encouragement to do whatever they can, as together, we can continue to build on a viable foundation for sustaining our urban trees and facilitating their long-term occupancy in our very dynamic world.

As we have seen, every aspect of Urban Forestry from planning and design to protection and preservation presents opportunities for citizen participation. From funding endeavors such as adopting trees or planting trees to celebrate or memorialize to global advocacy on behalf of urban trees in Sheffield UK or forest trees in Białowieża forest, Poland, citizens around the globe are finding all sorts of ways to embrace trees wherever they might be. Philosophically – it doesn't matter where the tree is – all trees matter. There are exciting, inspiring actions being taken all over the world thus, there is much to share and much to learn from all the varied ways in which we are connecting, all because of our trees.

Beyond Trees – Reconnecting with the Land

As I kneel to put the seed in
careful as stitching, I am in love.
You are the bed we all sleep on.
You are the food we eat, the food
we are, the food we will become.
We are walking trees rooted in you.
Marge Piercy

Not only is the Urban Forest inclusive, it's also expansive and adaptable. Although this is a book primarily about Tree →← People connections, we know that trees do not exist in a void. Their extensive community includes associates which they directly or indirectly support. And, more importantly, trees and their associates are connected through the land – the soil – our Mother Earth – and that underground web of life. It is the soil that facilitates all the connections and it is the soil that makes life possible. While my own personal predilection is trees and the landscapes they populate, as an Urban Forester I know that the Urban Forest functions at its best when people embrace its expansiveness and invest themselves in green endeavors which can re-establish our connection to the land – the soil - and bring us back into the circle, back to our roots. Community gardening and, more recently, urban agriculture and urban forest farming as well as community forestry, teaching urban forestry participation and participation through inspirational intangibles are all critical parts of that connective participatory menu which we will explore in Chap. 5.

Forest Doorways – Connections to the Universe

The clearest way into the Universe
is through a forest wilderness.
John Muir

Fig. 4.3 Massett-Teill forest reflection. Haida Gwaii. 2017. Naomi Zürcher

As we look at how trees have inspired connections, collaborations, relationships the world over, it seemed appropriate to explore the cycling/ recycling of the lives of forest and the Duwamish peoples, who have coexisted with their traditional forested lands for millennia. The most famous of the Duwamish (members of the Coast Salish community) was Chief Seattle (Sealth), known as an orator and a diplomat because of his eloquent statements in his effort to bridge the ever-widening gap between his people and Europeans. The City of Seattle was named after him.

The Duwamish, as with most indigenous cultures, place great importance on their connections with their ancestors. I learned of their extraordinary ancestral connection to the forests which were part of their traditional homeland along the west coast of the Duwamish River from the video "The Power of Trees". In this video, Ken Workman, Duwamish Council Member and great-great grandson of Chief Seattle, speaks to cultural traditions "... Up until a couple hundred years ago, when our people would pass, we would take them up to these very woods right here and that's where we would bury them. We'd bury them in the trees. The spring rains come and all that stuff on the ground, it goes back up into the trees so Grandma and Grandpa literally live in the trees. And so we are literally connected to this ground as a People into these trees. It's the cycle of life, like Grandpa said. Chief Seattle said that "we're just a strand in this web. We don't own the Earth, we are part of the Earth" and so it's very important for us to have healthy trees, to keep them growing, to maintain that connection as a person, as a human being, to my ancestors that are in the trees and to this ground that we now call Seattle."

All things share the same breath –
the beast, the tree, the man...
the air shares its spirit with all the life it supports.
Chief Seattle

Fig. 4.4 Tow Hill forest flora and fauna. Haida Gwaii. 2017. Naomi Zürcher

References

Fisher DR, Svendsen ES, Connolly J (2015) Urban environmental stewardship and civic engagement: how planting trees strengthens the roots of democracy. Routledge, London (4)

Koeser AK et al (2016) The cost of not maintaining trees: findings and recommendations from an international symposium and Summit Arboricult Urban For 42 (6):377–288 (4)

Mulcahy K, Zürcher N (2007) Tree shopping guide – selecting quality in the nursery.pdf. NYC Root Zone

Sommer R, Learey F, Summit J, Tirrell M (1994a) Social benefits of resident involvement in tree planting: comparison with developer-planted trees. J Arboric 20(6):323–328 (4)

Sommer R, Learey F, Summit J, Tirrell M (1994b) The social benefits of resident involvement in tree planting. J Arboric 20(3):170–175 (4)

Vogt J et al (2015) The costs of maintaining and not maintaining the urban Forest: a review of the urban forestry and arboriculture literature. Arboricult Urban For 41(6):293–322 (4)

Watkins SL (2017) Does collaborative tree planting between nonprofits and neighborhood groups improve neighborhood community capacity? Cities 74:83–99. https://doi.org/10.1016/j.cities.2017.11.006 (4)

Westphal LM (2003) Urban greening and social benefits: a study of empowerment outcomes. J Arboric 29(3):137–147 (4)

Zürcher N (2007) Noteworthy requirements for conifer and hardwood selection. NYC Root Zone

Zürcher N (2017) Assessing the ecosystem services deliverable: the critical role of the urban tree inventory. In: Pearlmutter D et al (eds) The urban forest. Future city, vol 7. Springer, Cham. https://doi.org/10.1007/978-3-319-50280-9_11

Chapter 5
The Global Experience in Green Participatory Place-Making: a worldwide sampling of Urban and Community Forest connectivity

An Urban and Community Forest odyssey, exploring examples of tree →← people connections which have been implemented on 6 continents

Keywords Citizens · Community · Connections · Degradation · Education · Goals · Mission · Outreach · Partnerships · Place-making · Poverty · Regeneration · Sustainable development · Stewardship · Visioning

As guardians of the good earth,
trees stand silent sentry over our planet,
strengthening our soil, creating the air we breathe,
enriching our lives with their natural splendor.
When we protect and nurture our trees,
we protect and nurture ourselves.
Mario Cuomo, Governor, State of New York.
Official statement from Governor Cuomo in honor of the publishing of the New York ReLeaf Information Kit and the initiation of the program.

Trees – social, sentient beings. As Suzanne Simard so eloquently offered in her TED presentation, dated August 2016: "…Trees talk!!! Through back and forth conversations, they increase the resilience of the whole community…" https://www.ted.com/talks/suzanne_simard_how_trees_talk_to_each_other?language=uk.

Trees are nurturers. They nurture each other and they also nurture us. It's not surprising, then, that so many of us have been touched by the warmth of such an embrace and have responded in kind.

The vast array of Steward actions on behalf of our trees leaves one breathless. It is quite clear that wherever trees exist, there is a tree →← people love affair in progress. Many people have told me *I've loved trees my whole life*. Loving trees usually equates with caring about trees, acknowledging their presence and, from my

Supplementary Information The online version contains supplementary material available at [https://doi.org/10.1007/978-3-030-94534-3_5].

N. Zürcher, *Connecting Trees with People*, Future City 16,
https://doi.org/10.1007/978-3-030-94534-3_5

perspective, that's an essential ingredient. But, if you're going to prepare a delectable dish, it will usually take more than one ingredient. Most dishes have something that binds them together and for trees, it's soil – soil as the basis for the entire community. In a world that is so diverse, with its vast array of Genera and species within Genera, climate and culture play an important role in who's doing what, with what and where that's happening. From such a world view, the participatory example brush had to be broad indeed and as the author, I never lost sight of that fundamental ingredient – the soil – as the premier nurturer, Mother Earth.

Aboriginal people and the land
… for the Aboriginal people land is a dynamic notion; it is something that is creative …
Land is the generation point of your existence …
it's the spirit from which Aboriginal existence comes.
It's a place; a living thing made up of sky, of clouds, of rivers, of trees, of the wind,
of the sand, and of the Spirit that's created all those things;
the Spirit that has planted my own spirit there, my own country …
It belongs to me; I belong to the land;
I rest in it; I come from there.
Patrick Lionel Djargun Dodson
"Report of the Third Annual Queensland Conference of the Aboriginal and Islanders Catholic Council of Australia", 1976, Page 16.

As this is a book about connections, so was the investigation and sampling of the participatory management repertoire. What is offered in the following pages struck both a professional and personal chord. Where a dialogue or an interview was possible, I have presented the thoughts of the project correspondent in their own words in the interest of connecting you, the Reader with their specific participation experience. The intent was to offer enough information – enough detail – to provide an in-depth understanding of exactly why and how in order to inspire the Reader and to make replication possible. Regardless of where the project has been realized, there are always elements that cross borders, arouse interest and motivate a *we could do that* thought process.

The programs that have been researched are offered topically, following a management plan order. The topic each program is assigned to is based on the primary focus of the program but many of these programs include additional management components. As I consider everyone a citizen and that all connections can make a contribution, I have also included professional *pro bono* participation as well as an array of participation intangibles that offer inspiration, an essential ingredient in citizen empowerment. In exploring the repertoire, I decided to preface the varied topics by offering an introduction to two locations which have initiated a multifaceted approach.

In my experience,
each country has a flavor all its own,
a mystique unique to it and to its people.
This flavor or "feel", is independent of topography,
flora and fauna, temperature, climate or even smells.
It springs from old nature spirits, gods, ghosts, witches, and demons,
which are a vital part of most peoples' worlds.
Edward Hall 1994.

Canada took the Urban & Community Forestry ball and ran with it, embracing Urban & Community Forestry at the country level, beginning with the Canadian Urban Forest Strategy initiative's opening statement: "The Canadian Urban Forest Strategy (CUFS) was first articulated in 2006 as a strategic initiative of Canada's urban forest practitioners including: foresters, managers, arborists, planners, community workers and politicians. It coincided with Canada's National Forest Strategy (1988–2008) a government, NGO and private sector coalition whose aim was to move Canada towards forest sustainability. The final iteration (2003–2008) of the National Forest Strategy included a section on urban forests."

Since 1992, Tree Canada has been implementing their vision of: "Improving the lives of Canadians by planting and nurturing trees while teaching about the value of trees." by planting more than 82 million trees, greening more than 660 schoolyards, helping to restore places hit by natural disasters and bringing together urban forestry experts, greening cities all across Canada. A crowning achievement has been the collaboration which produced the Canadian Urban Forest Strategy 2013–2018 (http://treecanada.ca/wp-content/uploads/2017/10/CUFN-RCFU-2013-2018-EN-Canadian-Urban-Forest-Strategy.pdf) and now 2019–2024 (https://treecanada.ca/wp-content/uploads/2019/04/TC-CUFS-2019-2024-Eng.pdf). A critical focus within the Strategy is Communication and Public Education which has been addressed through the outlined tasks: identify target audiences; develop a Communication Plan that establishes coordinated approaches to promoting the concepts of urban forestry to industry, government, educational institutions, and the public; develop communication *tools* to deliver the message; develop a funding strategy to support the public education program; produce educational kits for politicians, media, schools and community groups; create and maintain a Canadian Urban Forest Network website and the CANUFNET listserv.

This focus on the community as a critical stakeholder has enabled these tasks to be realized in Cities throughout Canada and the reader can find examples of some of these endeavors throughout the topical discussions.

Washington D.C. is not only the capital of the USA, it also carries the moniker "City of Trees". As with most Cities around the world, growth has been accompanied by a loss of tree canopy and that is no less true in the City of Trees. Growing D.C.'s canopy and resident well-being has been the focus, since 1982, of the City's premier nonprofit greening organization – Casey Trees. Their stated mission is "...to restore, enhance and protect the tree canopy of our nation's capital. To fulfill this mission, we plant trees; engage thousands of volunteers of all ages in tree planting, tree care and citizen science; provide year-round continuing education courses; monitor the city's tree canopy; develop interactive online tree tools; ensure that urban trees are reaching their full potential; work with elected officials, developers and community groups to protect and care for existing trees and to encourage the addition of new ones."

In this one mission statement, they have touched upon the depth and breadth – the very core – of citizen participation and their approach to realizing the mission will also be offered topically.

When Disaster Strikes

Disasters have the extraordinary potential to bring out the best in us. In times of crises, communities have a way of pulling together and lending a helping hand, especially in the face of natural disasters the likes of earthquakes, hurricanes and tornados. This has been no less true for disasters which have threatened our trees. It is tragic to see entire treed areas being devastated by some natural or unnatural occurrence but it does not have to remain a tragedy. The negative can lay the groundwork for a massive stakeholder collaboration enabling partnerships and empowered communities.

New York City Asian Longhorned Beetle Public Outreach Program

Citizen action was a very critical ingredient in the response to the discovery in 1996 of the Asian Longhorned Beetle (*Anaplophora glabripennis*) in the Greenpoint/Williamsburg community of Brooklyn NY.

The importance of citizens was evident from day one as the presence of ALB was not discovered by an Urban Forester or an Arborist or anyone working professionally in the urban landscape. It was discovered by a local resident – Mr. Ingram Carter – who was concerned that someone was vandalizing the street tree in front of his house by drilling holes along the trunk of this Norway Maple (*Acer platanoides*). Our local resident was so concerned with the damage that he decided to play detective to see if he could catch the *who* that was drilling these holes. To his astonishment, the culprit was not a *who* but a *what*. Not only was our local resident able to give the culprit a face (Fig. 5.1), his capture of the large and exquisite *what* was the opening salvo in a protracted war against Asian Longhorned Beetle – ALB – an invasive, very exotic, wood-boring insect that globalized its way from Asia to the US, Canada and Europe in untreated wooden pallets with a host list consisting of almost every urban-tolerant tree species. With thousands of infested trees being removed from NYC's Urban Forest, it became painfully clear that **this** problem could not be addressed without additional informed eyes.

Fig. 5.1 The Asian Longhorned Beetle (ALB). Kenneth R. Law, USDA APHIS

I proposed and was awarded Federal funding to develop an ALB Public Outreach Program for NYC's residents. The program intent: to inform all NYC residents about ALB and how they could help – the short-term, crisis-driven goal.

But also, to turn the disaster into a positive long-term benefit – to foster a real tree awareness and sense of stewardship. Rather than survey the entire City's trees for ALB – I was hoping instead, to grow the existing tree – people relationship – the one that already existed between people and the trees they are intimate with – to have them observe the tree that lives in front of the house where they live, the trees they pass every day on their way to work, or walking the dog or going to the supermarket. **Their** trees!!!

The extensive program that I had developed was devised to address as broad a base as possible. It's delivery was made possible with the help of a wonderful team of partners from Federal and State government agencies, not-for-profit institutions, local elected officials, and the private sector. The diverse program's highlights were:

Fig. 5.2 NYC Asian Longhorned Beetle Public Outreach Program components. Naomi Zürcher (**a**). Suggested Woody Substitutes; (**b**). Counter Top Display with Citizen Brochures; (**c**, **d**). Beetle Buster Trainee t-shirt; (**e**). Greenpoint Poster: Rally demanding replacement trees; (**f**). Beetle Buster Toolbox Host Herbarium with Patch

- Battling the Beetle: How to stop your neighborhood tree terminator – an all-day conference on the Asian Longhorned Beetle for professionals and citizens, funded by the USDA Forest Service Northeast Area, NYS Department of Environmental Conservation (NYS DEC) and the primary NYC utility company, Consolidated Edison Company of New York (Con Ed), as conference host. This all-day Saturday conference featured a morning of presentations from an international array of scientists, focused on providing an accessible understanding of ALB, its life cycle and management prospects in its non-native areas of infestation, followed by an extensive Q&A. The afternoon saw the 150 attendees (of which more than two thirds were from affected communities) divided into teams to search the constructed *infestation simulations* for signs of beetle presence. The subsidized registration fee of $15.00 covered attendance, all resource handouts, lunch + 2 coffee breaks and mini-grants were available for citizen attendees from disenfranchised communities. The idea was to introduce citizens to ALB, especially those living and working in ALB quarantine zones, and to the professionals and scientists who would be working with those community members as an accessible resource. It was the kick-off for all the Outreach Program components that followed.
- Suggested Woody Substitutes for ALB (Fig. 5.2a) – to open a dialogue and provide an awareness of ALB and to assist landscape professionals in choosing comparable alternatives to ALB host species. The resource presented a selection of host species, organized by the characteristics that made them traditionally distinctive and useful, grouped by color, texture, size, flowering and fruiting attributes as well as preference for growing conditions. What was then offered were urban-tolerant substitute non-host species as food-for-thought.
- Living & Working in an Asian Longhorned Beetle Quarantine Zone brochure. The brochure was developed as a useful tool with which to:

  - • identify ALB – where it came from, how it arrived in our communities and what to look for complete with ALB photos and descriptions of all its signs and symptoms;
  - • know what to do if there was evidence of ALB presence – either a suspected sighting of ALB or resulting tree damage;
  - • quarantine zone information and supporting maps.

The brochure was distributed:

  - • at all NYC/NYS tree/landscape-related conferences and workshops;
  - • in a countertop display, hosted by 100 local retailers (Fig. 5.2b);
  - • as a Con Edison utility bill "stuffer", mailed to every NYC customer.

- the Beetle Buster Toolbox – a traveling presentation kit containing script, slides, kiln-dried infested wood, a pinned ALB, larva, and frass. The Toolbox was developed by myself (donated services) in collaboration with the University of Vermont Entomology Lab with material contributions from Dr. Michael Smith of the USDA Agricultural Research Service along with a trainer t-shirt, designed by Alfred Zürcher (both donated services). The 25 Toolboxes were given to 25 greening organizations, committed to being trained in order to offer free workshops to communities within the quarantine zones. (In addition to funding the production of the trainer t-shirt (Fig. 5.2 c, d), Con Edison privately funded an additional 25 Toolboxes for interested public schools.) The Train the Trainer sessions were conducted by myself along with USDA Forest Service personnel at donated locations in Brooklyn, Queens and Manhattan.

Since the study of invasive species was part of required NY State curriculum, many public schools took advantage of the presence of an iconic exotic invasive, As a result, I had an opportunity to present many Toolbox-based lectures in schools around NY State, as ALB was an excellent example of an exotic invasive species with potentially devastating impacts on the entire State's environment and economy.

• Specifications for the Replacement Tree Planting Initiative. A Community Board 1 Re-greening/Planning Greenpoint/Williamsburg committee was formed in response to community protests (Fig. 5.2e) demanding replacement trees and included Sharon Ossenbruggen and myself. The goal was to develop a Strategic Plan for the re-greening of those areas devastated by ALB. This Plan would address the state of tree planting specifications with an interest in incorporating a more up-to-date Best Management Practices (BMP) approach to urban tree selection, planting and establishment.

The local professional arboricultural community along with the professional members of the NY Releaf Region 2 Planning Committee attended these evening meetings, *pro bono*, to produce a specification draft which would put forward an enhanced approach to the replacement initiative. The resulting specifications were submitted to the NYC Department of Parks & Recreation. The planning body also provided *pro bono* oversight of the replacement initiative to ensure the BMP specification deliverable.

• A Beetle Buster Patch Program (Fig. 5.2f) for Brownie, Junior, Cadette and Senior Girl Scout Troops – earned with Beetle Buster training, design and assembling of 25 ALB Preferred Host Herbaria for the 25 original Toolbox recipients plus ALB awareness activities in quarantined communities.

The Patch Program component was realized entirely through donated services and it was the partnership that made the deliverable possible:

- • given that the quarantine was in both Brooklyn and Queens, Brooklyn and Queens Botanical Gardens were approached and offered host species materials, classroom space and herbarium supplies;
- • the necessary complement of educators needed to train and work with participating Girl Scouts as they constructed the 25 herbaria and conducted ALB awareness activities were from the New York ReLeaf Region 2 Planning Committee and USDA Forest Service personnel;
- • the Girl Scout patch program administrator allowed me to design a unique patch which featured an Asian Longhorned Beetle.

- The "Greenpoint Counts Trees Because Trees Count" tree inventory. The purpose of the inventory was to determine what tree species still existed within the Greenpoint Community, public and private, how many of each species, their condition (professionally rated) as well as the number of available planting spaces for the replacement planting initiative.

A pre-inventory training module for the interested community was developed by myself in collaboration with our wonderful USDA Forest Service Urban Forester, Sharon Ossenbruggen. Training was provided by both Sharon and I and facilitated in the field by additional USDA Forest Service personnel along with concerned professionals from the local professional arboricultural community. Materials such as printed Excel-based spreadsheets, DBH tapes and a project T-Shirt for all participants was funded by the USDA Forest Service.

The **ALB Public Outreach Program** and its very modest funding trained over three thousand people and, through its varied activities, connected directly with tens of thousands of residents. The Con Edison bill stuffer mailing reached millions of NYC dwellers, all through donated services. People retained the ALB brochure for years after ALB was no longer in the headlines and would often call me with tree concerns or questions.

An important outcome of the Community Board 1 Strategic Plan, referenced above, was the development of a community-based educational endeavor including school curriculum. In addition, the partnerships and informed citizen connections that developed formed a basis for a collaboration that still exists today, realized through advocacy and taking ownership of our trees in their time of need. The Greenpoint community has exemplified that spirit of grassroots stewardship through the model Greenpoint Tree Corps program, presented under the section on Post Planting Care and Long Term Maintenance.

Assessing the Resource – Tree Inventories and i-Tree Assessments

Puerto Rico – The Community Response to Hurricane-Instigated Landscape Devastation

Puerto Rico, officially an unincorporated territory of the US, is part of the Greater Antilles archipelago located in the Caribbean. Its southern and western coastlines are on the Caribbean Sea and its shores to the north and east, the Atlantic Ocean.

September 2017 saw the arrival of two deadly hurricanes within a 2 week span – first, Irma, a category 5 and then Maria, a category 4. Maria's sustained hurricane force winds of 155 mph devastated the island's man-made infrastructure along with communications and utility services. In addition to the loss of man-made structures and services, the island's lush tropical vegetation took an extraordinary beating from extensive flooding and landslides.

Losses to natural land cover, especially in the form of forests, prompted post-storm assessments to determine the extent of the damage and the resulting loss to Ecosystem Services benefits. "A post-hurricane rapid assessment of island-wide green infrastructure using spectral mixture analysis (SMA) suggested 23–31 million trees may have been severely damaged or killed in Puerto Rico" (Feng et al. 2018). Estimates are needed describing how these damages translate into ecosystem services losses. Also needed are finer scale and on-the-ground measurements that provide information about the spatial variability of tree cover loss and the effect on changes in ecosystem services, in order to guide recovery efforts. Evaluating these services and their monetary benefits can facilitate Urban Forest recovery management and long term urban planning to achieve community resilience through disaster preparedness, mitigation and response." (Elvia Meléndez-Ackerman, PhD, et al. Ecological vulnerability of urban green infrastructure to Hurricanes Irma and Maria in Puerto Rico. A report to IITF US Forest Service as Part of the Post-Hurricane Assessment Campaign [Mar–May 2018]).

Given the need for resource assessment and the quantification of ecosystem services and their monetary value, Elvia and colleagues decided to implement an i-Tree Eco project to ascertain the needed information (Fig. 5.3).

Fig. 5.3 Puerto Rico Investigación i-Tree Poster. Elvia Meléndez-Ackerman

Elvia is a plant ecologist by training and an interdisciplinary researcher by experience which has expanded to address sustainability and adaptive capacity issues within the context of urban social-ecological systems. A focus on how feedbacks between social and ecological factors influence human well-being has resulted in collaborative team initiatives, implementing survey research that combines social and ecological data collection. In addition to serving as Principal Investigator (PI) on numerous federal and state agency-funded grants, Elvia worked as Co-PI on the NSF-funded project which involved the collaboration of highly interdisciplinary teams (planning, public health, ecology, sociology, architecture) across multiple institutions (academic and non- academic). Her lab – University of Puerto Rico at Rio Piedras (UPRRP) – has provided research experiences to over 120 students and has advised 5PhD and 4MSc in Biology and Environmental Sciences and generated research with local and international collaborators.

Elvia has kindly shared her post-hurricane Urban & Community Forestry assessment experiences with us, as follows:

NZ – Did anyone from your Lab or prospective team members receive I-Tree Eco training to facilitate your working with citizens?

***EMA* –** With support from the Forest Service, Dr. David Nowak provided a college-level course on forest inventories and i-Tree tools. The course helped train students and professionals from Puerto Rico and the Dominican Republic. A number of students in my lab (graduate and undergraduates) developed their research projects on the evaluation of Urban Forest resources in different sectors and neighborhoods in San Juan. Citizen participation has been crucial for the development of those inventories. The project team consisted of myself along with Ana Trujillo, Dr. Sofia Olivero and Dr. Chris Nytch. Ana's work focused on building structured participatory models of forest inventory that incorporated i-Tree Eco in San Juan's Santurce neighborhoods that have varied socio-economic profiles. Sofia's work focused on peoples yards in San Juan and Chris runs an ongoing citizen science project within the El Yunque National Forest.

After the hurricane, the International Institute for Tropical Forestry (IITF) funded a proposed post-hurricane i-Tree Eco/Canopy assessment to estimate ecosystem services loses using the data that Ana and Sofia had collected. The funds helped support five graduate students (Chris Nytch, Molly Ramsey, Benjamin Branoff, Evelyn Baez and Ana Trujillo) and 9 undergraduates. The proposal was funded by the Department of Interior (DOI) and the report was submitted to them as well as IITF. (https://www.researchgate.net/publication/327235826_Ecological_vulnerability_of_urban_green_infrastructure_to_Hurricanes_Irma_and_Maria_in_Puerto_Rico).

Reforestation following the hurricanes has been led by local NGOS (Para La Naturaleza) and San Juan Bay Estuary Program, both of which have volunteer programs. The complexity of social-economic problems have been so severe (inadequate post-hurricane response, promised aid that never came, emigration, political instability), I am afraid that green infrastructure recovery has been slow and scattered. When it comes to citizen involvement at the inventorying stage, Sofia, Chris and Ana (from our lab) were directly involved with the post-hurricane inventories and collected many anecdotes about what people did to keep their vegetation.

Pending is a Department of Natural Resources and the Environment proposal that targeted urban restoration projects and education with community involvement.

In terms of the recovery of urban trees, it seems that even with assessments, ownership was an issue. We have inventories for street trees in two urban developments (pre-post hurricane) with the post hurricane one using i-Tree. While we were able to document losses, some trees where in public places, some where in private places so even if there where funds available for tree recovery it was not clear who could make the claim. Needless to say that if there were funds available, that process was not clear either.

NZ – You mentioned the college level course that Dave Nowak provided. Did the students and professionals that were trained go on to work with the participatory projects that were part of your lab?

***EMA* –** Yes, the course trained professionals from the Dominican Republic and Puerto Rico as well as students (graduates and undergraduates) from the University of Puerto Rico (UPR) at Rio Piedras.

In 2015, David came to UPR. Training was possible through USAID (Aid for International Development) funding to the USDA Forest Service and allowed the group from the Dominican Republic to come to Puerto Rico and for David to come as well. Gerald Bauer (Jerry) from IITF was the main facilitator. In 2016, David returned to Puerto Rico to do a repeat of the course, which was funded by UPR. Below, I describe how the i-Tree-course alumni integrated into different activities sponsored by the lab. I carried out a modified version of the course in the spring of 2018 (after the hurricanes).

The group from the Dominican Republic included personnel from NGOs, University, Municipal Government (National District of Santo Domingo). That led to collaborations funded through US Aid for International Development (USAID) and Partnerships for Enhanced Engagement in Research [PEER] (https://www.nsf.gov/funding/pgm_summ.jsp?pims_id=504726) which allowed us to do i-Tree inventories in different areas of the Santo Domingo National district and several college campuses in Hispaniola. The model was to use alumni from the Puerto Rico (PR) course as trainers in various Train-the-Trainer workshops that then resulted in the training of students and faculty from the Dominican Republic (DR). Mervin Perez, a Guatemalan graduate of UPR, was the leader and primary resource and, as a result, he has done trainings in other cities in Central America. NGO and Municipal staff from DR helped with the participant recruitment in Santo Domingo and facilitated ancillary data collections and agreements within the Dominican Republic. Two techs, contracted through the PEER project, included an i-Tree component in their undergrad thesis and are currently in grad school at INTEC. Another, Ana Pou, was an alumni of the first i-Tree course and now has a state government job in management.

In Puerto Rico, both Ana and Sofia used i-Tree for their dissertation work and developed a general principle for student recruitment: providing training to a target audience that would then be recruited to participate in later inventories. Ana facilitated a collaboration with the NGO Foundation for Puerto Rico and UPRRP – my lab – to provide training to communities in Santurce. Through that collaboration, Foundation for Puerto Rico wrote an Americorps proposal that in turn helped support the training of undergraduates who would then serve as resources for the community trainings in i-Tree Eco and the i-Tree assessments.

Sofia evaluated ecosystem services in residential yards and also trained students to help her with her work. Following the hurricane, she submitted an Experiment.com project that was funded and allowed her to finish her dissertation data collection and to fund herself as well as other students.

Chris is also an i-Tree alumni, working as an independent contractor and the PI of a grant submitted to the Forest Service via Amigos del Yunque (a local NGO) and to whom my lab offers facilities support. Chris along with the project-funded staff have followed a similar model to Ana's in working with the community but with some modifications given the scope of the project.

Our lab trained two teachers at a Rio Grande Mid-School in i-Tree and facilitated an i-Tree assessment before the hurricane in 2016. After that hurricane, they were able to document tree mortality which was quite high (approximately 60%).

Last, at the time of the hurricane, I had just started a project that focused on forest inventories with a Forest health evaluation component. This project allowed us to conduct i-Tree assessments in several areas and provided research opportunities to students. I also mentored additional students through other i-Tree projects (Gabriel Rodriguez, Stephanie Rodriguez, Jose Garcia – all now in graduate school).

NZ – Has the work that the NGOs were/are doing building on the inventories/I-Tree assessments that your lab did? Were they the basis for the initiatives the NGOs undertook?

EMA – The NGOs at Puerto Rico (Para La naturaleza, San Juan Bay Estuary program) that are leading tree plantings are not necessarily using the inventories/assessment data to decide where to plant but rather to justify the need to replant.

Studies that came out after the hurricane on the loss of trees (from journals and technical reports, including the post-hurricane i-Tree report) were used to justify the need for planting. They provided technical expertise and tree materials to organizations that request tree planting and there is a commitment from organizations to take care of the planting. I believe that this is ongoing but not much publicity has been given to this. I am also not sure how widespread this is either as it is not publicly discussed.

NZ – Is any official entity providing replanting stock as part of the recovery? Is that FEMA? If the intent is recovery and the documentation of loss is evidenced by a recognized valuation assessment strategy, why has this been encumbered by an unclear process? Normally, replacement is for both public and private as the loss to the general public is much greater than to private property value?

EMA – In my discussions with Sofia, the lack of official inventories (or track of them), the lack of maintenance and follow up of tree resources, and lack of a comprehensive tree management plan then precludes local government from claiming the loss of particular tree resources. That being said, the residential space in the San Juan Metro Area is about 30% so lots of trees in private hands. Also, I am not sure that public or private stakeholders understand how FEMA or any agency may help recover the loss of forest resources (this is definitely a new concept for private citizens and maybe for the local government).

i-Tree specifics

In order to follow up on the i-Tree component of this Puerto Rico/USDA Forest Service post hurricane endeavor, I did an outreach to Dr. Chris Nytch who, as noted above, personally participated in the implementation of i-Tree in the El Yunque National Forest. While Chris was directly involved in the use of i-Tree Eco and Canopy to assess the loss of Ecosystem services as a result of the two hurricanes, he has since helped to initiate a Citizen Science vegetation monitoring project in the secondary forested lands of El Yunque – Redescubre El Yunque/Rediscover El Yunque (https://elyunque.org). This project is an outcome of the Revised Land Management Plan, developed by the USDA Forest Service in a "shared stewardship" collaboration with the Public, stakeholders and forest employees to "Maintain a healthy, accessible, and sustainable forest that integrates multiple uses; provides economic, ecological, and social opportunities; promotes education, environmental justice, cultural and environmental identity and awareness for the conservation of its natural resources; and adaptive forest management that is inclusive and collaborative".

What do we do?

- Help to improve and maintain the quality of life of the Forest and surrounding communities.
- Provide and promote access to the Forest and the sustainably of its resources in collaboration with the multiple socioeconomic sectors of the public that benefit from it
- Ensure access to scientific and interpretive information available about the Forest, including its cultural resources and Forest services and products.

How will we do this?

- Collaborating in sustainable socioeconomic development and community activities that help in conserving and improving the Forest and its resources.

• Developing sustainable and collaborative projects with the public and private sectors and society at large.
• Conserving, preserving, and restoring Forest habitats and ecosystems through strategic adaptive planning, research, monitoring, and education.
• Promoting and strengthening the shared stewardship and the social and ecological identity and connectivity of the region through improved public participation and involvement in decision-making.
• Using the best science and local knowledge.

For whom will we do this?

• For the welfare of nature and society based on a local focus and global perspectives." (https://www.fs.usda.gov/Internet/FSE_DOCUMENTS/fseprd592253.pdf)
Chris has provided details for this initiative as follows:

> **NZ –** Which entity took the initiative to develop the program? I am assuming that the Fundacíon Amigos de El Yunque (FADEY) is a primary funder but were they also the initiators or did this come from the communities or one of the other NGOs or from the USDA Forest Service? Through which partner are you participating?

CN – The USDA Forest Service (FS) is the primary funder. The citizen science idea stemmed from a joint interest by the Forest Service and FADEY to develop a volunteer program that emphasizes scientific monitoring and community engagement, both of which are primary objectives of the forest's revised management plan. Following Hurricane María, FADEY approached FS with a specific project focus on vegetation monitoring within a management area of El Yunque known as the Community Interface Resource Management Area, which is comprised of 2907 ha of secondary forest lands located at lower elevations (<600 m) and adjacent to private landholdings. I work for FADEY as the Project Leader/Lead Scientist and developed the proposal with colleagues which, once accepted, was formalized in a collaborative agreement that began in October 2018. The project will end at the end of September 2020.

> **NZ –** I see there is a volunteer form provided at the Redescubre El Yunque website (https://elyunque.org) but through what additional methods and mechanisms is the project doing outreach to potential participants?

CN – Outreach is shared among three project partners. FADEY does outreach through social media outlets such as Facebook and Twitter, as well as in-person tabling activities or talks at environmental fairs, universities, and also at some schools. We also do outreach by direct personal contact with citizens, civic groups, and leaders from local communities around El Yunque with whom we have previous contact from other activities, as well as with people we've met along the way while conducting training workshops (more on this below) and while working in the field.

Additional outreach is conducted by the NGO Corazón Latino. Via a separate collaborative agreement with the FS, Corazón Latino developed a media strategy for communicating broadly about the vegetation monitoring project and providing effective branding (development of Redescubre El Yunque logo, design of elyunque.org landing page, design of informational flyers). Within the past year, they have also contracted a person locally to help with on-the-ground connections with communities around El Yunque. The third outreach partner is the FS, which plays a key role in volunteer coordination/communication with citizens. Specifically, there is a FS intern who is responsible for serving as a liaison between myself/my science team and the population of interested and already committed volunteers. He provides follow up with our volunteer community about upcoming workshops and field workdays, and deals with the logistics to get people to the activities.

NZ – Are there any criteria – age, etc. – or just an interest in getting involved?

CN – As you note, the main criterion is an interest in getting involved. We accept volunteers aged 5 and up. The fieldwork can have some demanding physical challenges, such as steep slopes, rain and mud, dense vegetation, bugs and some plants that can cause allergic reactions, but we let volunteers decide for themselves what they are capable of/up for. We also accept volunteer support in the capacity of helping to do outreach to potential volunteers in the local community and with computer data entry.

NZ – How many citizens have participated to date? Are there any demographics, e.g. age, gender, etc.?

CN – To date, and since the project began outreach to volunteers in March 2019, we have engaged with 162 citizens in one form or another. This total includes people who have come one time for a full or abbreviated training workshop that involved a combination of conceptual background knowledge and hands-on training using adapted i-Tree ECO methodology, and/or have participated in field data collection. It also includes both residents of PR and some visiting guests from the US (<10%) who were connected with the project through the partner organizations. Of the 162 citizens engaged, 66 have participated in field data collection at one or more monitoring plots. We do not collect information about age, gender and other demographic characteristics, per FS guidelines, other than to request that minors have written permission from a parent/guardian, and that kids under 12 be accompanied by an adult. My informal assessment based on visual observation is that a majority of participants fall in the young adult age range of 18–30 years old, and roughly 2/3 are female.

NZ – Once volunteers have been recruited, what sort of training will they receive and from whom? How have the trainers been trained? If the trainers are University students, does participation in this project count towards study credit?

CN – Volunteers are trained by a project manager and a team of 6–9 assistants who work directly with me. I trained assistants based on i-Tree methodology that I learned from David Nowak and colleagues in cooperation with Elvia Meléndez and partners in Puerto Rico who use i-Tree. I also participated in the i-Tree Global Users Symposium in 2019 which provided an opportunity to refine the methodology for tropical settings. There is continual interaction between myself and the project manager and assistants on a weekly basis.

Several (but not all) of the assistants are undergraduate students and hence the range of 6–9 individuals because the number varies per semester depending on availability. Currently, we have 7 assistants who work between 10 and 55 h per month. The team of assistants has included people who receive undergraduate research credits (which requires work additional to the project to meet the course requirements), or an hourly wage, or who participate voluntarily in an unpaid research position with FADEY (because they were interested in gaining field experience but for some reason not eligible for credit or financial compensation). Some assistants started as volunteers and then transitioned into a paid position. The element of providing financial compensation for those who want/need it has been key in maintaining a consistent core team of qualified assistants throughout the project. This consistency has benefited their personal professional development, as well as the quality of the data collection and the experience of volunteer participants.

Most training of volunteers has occurred in four community workshops in municipalities around El Yunque over the past 9 months. The workshops have been held at local community centers and the number of participants has ranged from 25 to 55 per workshop. The workshop consists of a morning session dedicated to explaining the project context and current forest status, and introducing the concept of ecosystem services and the role of trees in providing benefits to human communities. This is followed by lunch and then an afternoon practice session that introduces the tools and methodology used in the field. We close with a round of certificates and a call to join us on workdays in the field. We opted for this training method as a means to present potential volunteers with the necessary information in a controlled environment before going to the field, both for their learning benefit and to increase efficiency of data collection in the field during workdays. We have also conducted several mini-training sessions with select groups of interested volunteers, which consist of a condensed version of the content of the full community workshops.

Volunteers participate in the field every other Saturday and every other Wednesday. This provides availability during the week and on weekends. A typical field day for volunteers is from 8 am to 12:30 pm. Alternating one week on, one week off gives our team of assistants time to scout and establish new plots, do field checks, maintain equipment, and enter data.

> **NZ –** Is there a curriculum? Does any of this information exist online and would you be willing to share the details given that the intent of this is to inspire readers and encourage replication?

CN – There is a curriculum. It is based on a modified i-Tree Eco methodology. Modifications include adaptations to working in a forested environment as opposed to an urban one, and the incorporation of elements relevant to the humid tropics (e.g., extensive palm trees and lianas) and of interest to the management El Yunque (e.g., native vs. non-native species, hurricane damage). The curriculum exists primarily as a series of PowerPoint presentations (used in the training workshops) that

explain the field procedures. We also have an informational booklet, some promotional videos – see these links:
https://drive.google.com/open?id=1y-NuZdZOL7sSqCgeSNr3ZTWtHKSWseMw
https://drive.google.com/open?id=1W5MQfsq3AR_bGz3A3YjAt8EqTTx0GoOE

With the exception of the booklet and the videos we have not posted the information online primarily because we have not had the time/capacity to adapt the other resources to a web environment. We are currently developing a digital atlas that will summarize/highlight the purpose/context of the project, volunteer participation, vegetation data, and ecosystem services. This will be ready by the end of September 2020 and will include links to i-Tree resources. I do not think we will have time/resources to include more detailed training curriculum, as this was not part of the original agreement with the FS.

NZ – I-Tree Eco field data collection can be somewhat complicated for people who have never had to do a tree inventory. How has the training dealt with these complications? Are there always experts in the field accompanying the volunteers? What aspects of the data collection require the greatest *hand holding*?

***CN* –** Yes, it can be challenging. We deal with this by always having experts (assistants, project manager, or me) in the field with volunteers. The model of having a consistent team of assistants who work directly with volunteers in the field allows for a quality learning experience for volunteers (typically a ratio of 1 or 2 volunteers per assistant). Likewise, when no volunteers come to the field on a particular day, having a team of assistants allows us to continue with the project objective of collecting data from the 40 monitoring plots. The part that is most challenging are the subjective decisions/analyses about where to take the DBH (because of uneven ground, or irregularities of the tree bole) or how to estimate percent missing and dieback. We also review the data for anomalies and include field checks for data points that are missing or seem erroneous.

NZ – Has the project extended its work into neighboring schools? Has any of the *why* or *what* been incorporated into school curriculum? Are there school outings to El Yunque where the students can see the areas where an I-Tree Eco assessment has been done so that the resulting data takes on greater relevance?

***CN* –** We are interested in extending into neighboring schools, and have made approaches to a few local ones. Likewise, some schools have approached us as well. However, the initial focus was not directed toward schools in particular. Given the dense post-hurricane vegetation conditions, we cannot safely accommodate more than 10–12 volunteers at a time, and the average class is usually much larger. Furthermore, resources were not built into the budget to facilitate the participation of large school groups (e.g., transportation to the field). We have addressed this issue in part by inviting interested teachers to participate with a small group of dedicated students who can learn the methodology and then share it with the rest of their class. We have also written a proposal for additional funding to support the direct training of several schools in municipalities around El Yunque. The idea is that we would provide directed training of i-Tree Eco methods in their schoolyards first, after which we would invite them to come to the forest in teams over several days to help work in the monitoring plots.

Through our partnerships with educational programs that run parallel to ours (such as the Luquillo Long-Term Ecological Research – Schoolyard Program) we have been able to share some of the El Yunque and i-Tree project content with teachers who are interested, but other than that, the project has not yet been formally incorporated into school curriculum. Nor do we have school outings to El Yunque as part of the project where students can see areas where i-Tree Eco assessment has been conducted. We would definitely like to do more of that and help the results take on a greater relevance. That will likely have to happen as part of future projects.

NZ – I-Tree Eco requires the use of certain equipment – a DBH tape, a clinometer, etc. Is this equipment donated or is there funding for its purchase?

CN **–** Most of the equipment was purchased with funds provided by the FS. A small fraction has been donated by partners, including safety equipment and some field tools.

NZ – Once the I-Tree Eco field data has been collected, are citizens also inputting the data or is the Forest Service or another partner completing this task?

CN **–** Most of the data input is being done by the same assistants who guide volunteers in the field. There are a small number of volunteers who have also helped with data entry.

I will note that because of several differences in our tropical forest-based methodology, and also because of the management objectives of the FS, we created a database apart from i-Tree for archiving of data, which includes both tree data as well as data about volunteer participation. For the upcoming analyses of ecosystem services we will migrate/import the necessary parameters to the i-Tree software.

NZ – How does the project retain the connection to volunteers? Does the project continue to engage the participating volunteers in other tasks? If not, do you try to connect participants to other such projects?

CN **–** As mentioned above, FADEY keeps in touch with volunteers and there is also a FS intern who serves as a liaison and communicates with volunteers about upcoming activities including workshops and field workdays. Via our social media outlets, we are often communicating passively to volunteers about other projects and occasionally we do direct outreach to individuals as well. Most volunteers have come to the field once or twice. There is a handful of super volunteers who have come five or more times and have gotten quite good at the data collection, to the point that I would trust them to lead other volunteers. We've recognized a few of them with certificates highlighting their accomplishments.

Fundacîon Amigos de El Yunque http://www.amigosdelyunque.org

UK – i-Tree Canopy Cover Map

As we strive to connect more and more people to how we can address Climate Change impacts through actionable initiatives, the i-Tree suite of freeware assessment programs continues to be an important source of data that enables us to quantify the benefits our trees are delivering. Not only do these programs provide scientifically documented data output which can be shared far and wide, they also afford the grassroots the ability to get involved by participating in the collection of the data required for input.

i-Tree's implementation at the grassroots level is nowhere more evident than in the UK where numerous communities have been involved with i-Tree Eco assessments. A new initiative, entitled the Urban Tree Canopy Cover Map, had been created to find out exactly how much live canopy exists at the ward (local) level in order to facilitate a greater understanding and management of the resource.

This initiative has taken a broad brush *anyone who wants to take part* approach to participating – professionals, citizens, students… Data collection from each ward is implemented through the use of the online tool i-Tree Canopy https://canopy.itreetools.org/ and https://canopy.itreetools.org/resources/iTree_Canopy_Methodology.pdf.

Instructions are provided through an online link, https://www.forestresearch.gov.uk/research/i-tree-eco/urbancanopycover/

and the data that's collected is added to an online interactive map http://forestry.maps.arcgis.com/apps/webappviewer/index.html?id=d8c253ab17e1412586d9774d1a09fa07

Given the potential for the information that has been collected, it's important to know how the data might be used, beyond the interactive access. The following questions were posed to Kieron Doick, head of the Urban Forest Research Group at Forest Research UK:

NZ – For the non-UK reader, can you very briefly explain what a ward is?
KD – A ward is the primary unit in the English electoral geography for civic parishes, boroughs, and district councils. Or more simply, it is 'a local authority area (sub unit), typically used for electoral purposes'. (we use it because citizens can relate to it and it's potentially a way to spark political debate, as wards are represented by Councillors)

NZ – How many wards are included in your project? Is it the entire UK?
KD – Actually it's not quite that simple. We have 5091 'urban' wards visible on the basic map. We wanted to make all (rural and urban; in fact ALL the wards of the UK) wards available. However, we wanted to focus people/initial efforts on the urban wards (as that's the focus of the Urban Forest Research Group). To do this, we used local authority classifications, screened, and 'urban' wards selected by population density. We then made them visible on the map. (People can access the other wards by changing a simple setting in the menu, and then all wards will show).

NZ – What inspired the project's creation?
KD – In 2016, we conducted a study to assess urban canopy cover across ~300 towns and cities in England. We also did seven Scottish cities, and Cardiff in Wales. We presented the work in 2017 at the Institute of Chartered Forester's '3rd Trees People and the Built Environment' International Conference. The data was well received. But, it was too course grain to drive action, and it would have been too costly to use the same approach for ~5000 wards – a ward takes about the same time to survey as a regular sized town or city as both are assessed using 500 standard sampling plots. So, we looked for a way of taking the project out to volunteers, citizen scientists. The major delay was finding a means of getting the required GIS 'shape file' to people to use in i-Tree Canopy. It is too difficult to expect people to 'draw' their ward within i-Tree Canopy, and very challenging to guide people in how to source the file from, for example, an internet search. We had to make it as easy as we could, and arguably we are still a bit too complicated by having one website to learn about the project, a second to access the Shape files and view the project's results, and a third (admittedly 3rd party) site where the actual analysis takes place.

NZ – Besides the posting in LinkedIn, how are you promoting participation? Have you

- done an outreach to known environmental or greening groups as well as those indirectly related, e.g. birding groups,

KD – We partnered with the tree charity Trees for Cities to design our User Guide and launch the project and they have promoted the project throughout their membership. We have presented the project to the Woodland Trust, who have shared it with their membership (through dissemination by their Partnership officers). We have also presented to the Tree Council, and, on seminar days, to the Tree Council's Tree Wardens

- gotten articles or announcements into major and minor newspapers,

KD – While we haven't targeted regular media, we have submitted articles to ArbMagazine, and Pro-Landscaper, both trade journals, the first to Arborists, the second to the landscaping/horticultural trade.

- used social media,

KD – Featured in Tweets rather a lot, especially at the beginning

- approached schools to try to incorporate as an experiential extension to curriculum,

KD – Not schools other than one secondary school child tested the User Guide for us before launch; a couple of Universities have used it within Tutorials, and also presented as part of a wider talk to a university Environmental Science department on the benefits of urban trees.

- other – please explain?

KD – We've presented at a range of conferences and other meetings; including Research update meetings to the Forestry Commission, and their 'Forest and Woodland Advisory Committees' (statutory advisory meetings that meeting regionally across England)

NZ – Do you have any idea of how many people are reporting data?
KD – Only 77 users to date (that's in one year); 31 of which were still actively supplying data during the past 3-months

NZ – Through what mechanisms are you trying to involve professionals in the data collection?
KD – We have presented at the Arboricultural Associations annual conference in Sept 2019. The project was also featured in Forest Research's Newsletter on a couple of occasions

NZ – What are some of the ways you are trying or planning to raise project awareness and encourage incorporation of the data into Urban Forest management?
KD – By conducting outreach as widely as possible, and an ongoing effort to look for new ways to reach new audiences, we expect the data to become known to those who might have an interest and/or who need it – such as local Councillors and local authority Tree Officers. We are also reaching out to the Tree Officer Organisations, and Local Nature Partnerships (who will have local authority representation). Through these varied initiatives, we are sure that the right people will learn of the data. How they use the data is then up to them however, we have been involved in a number of projects commissioned by a local authority to measure canopy cover at ward level and by promoting these projects we have some confidence that some potential data users will be aware of examples of how this sort of tree data can be used to support local tree planning and planting. While we don't really have 'municipal urban foresters' in the UK, local authorities do have tree officers and green infrastructure officers, and our strategy has been to promote the project (and it's data) directly to them

NZ – Any other information you feel might be relevant would also be appreciated?
KD – No, but your questions have certainly helped me think about where we have and haven't worked hard enough to disseminate and get more people on board. This project started off small with very little funding, but I'm going to aim to give it more focus next year.

Unley, South Australia – Community of Possibilities – connecting the data output to the community with i-Tree Eco assessments

The City of Unley, South Australia began its *Community of Possibilities* initiative in June and July of 2012. The initiative engaged with 1200 community people in order to gather ideas for Unley's Environmental Sustainability Strategy. One of the outcomes of this endeavor was the formulation of a detailed City of Unley Tree Strategy (Fig. 5.4).

Fig. 5.4 City of Unley Australia i-Tree assessment/Tree Tag project. Jenni Garden

As a result of discussions between Dr. Jenni Garden, representing the contracted private Consulting company, Seed Consulting Services and the City of Unley's tree strategists and in response to the City's interest in valuing its trees, it was decided to do an i-Tree Eco assessment of the City's largest Park, Ridge Park.

Dr. Jenni Garden is a private Consultant who specializes in urban, landscape and climate change ecology, providing science-based innovative solutions to create more livable cities for people and the environment. Her greatest challenge is creating a better balance between environmental, urban development, and social health and well-being requirements. In addressing that challenge, Jenni has kindly provided details on the Tree Tag project, as follows:

NZ – How did the i-Tree assessment reports prompt the Tree Tag idea? How did you create and initiate the model?

***JG* –** Tree Tags came about from a discussion with a local Council Client (City of Unley) as to how to engage the community about tree values determined from the i-Tree Eco assessment. Given that the i-Tree Eco outputs are quite technical, we had to come up with a way to make the data more accessible.

I had seen examples of large novelty tree tags used in various locations in the US which told people the dollar value of ecosystem services provided by the tree. I took this idea to Council as a potential approach for engaging the community here. The Council loved the idea but wanted to avoid putting a dollar value on the tree – trying to avoid reducing the tree to a monetary figure. Instead, through a brainstorming process, we decided we could provide information on the actual services provided but convert them into relatable metrics – the first suggestion was of the number of beach umbrellas instead of square meters of shade….and things evolved from there.

The City of Unley designed the first tree tags and we launched them at a community Nature Play event and did some surveys of community members to get feedback on the tree tags – the report for this is also on the i-Tree website. Since these first Tree Tags, the messaging and Tag design have evolved for a number of other local council clients around Australia – using information and relatable metrics of most interest to them and their community.

NZ – I saw that you have schools participating not just in Southern Australia, where you are located, but in Queensland as well. How have you communicated the Tree Tag project concept with school districts where you are based and beyond? Is there literature that you have provided or information through local media or on a website that shows the basics and how to get involved?

***JG* –** Regarding the Tree Tags in schools, communicating this within and beyond South Australia has been a combination of: (1) meetings with potential clients and letting them know about services we offer, including Tree Tags, (2) word of mouth and people seeing the Tree Tags hanging in different Councils, and (3) presentations at conferences

NZ – What is the age range of the students who are presently participating?

***JG* –** So far we've run Tree Tag projects with:

- Year 5 ESL class, ages 9–10 (South Australia)
- Day-care/preschool group, ages 3–4 (South Australia)
- Year 6/7 class, ages 10–11 (South Australia)

- A range of years from late primary to early high school (Queensland) – this was part of the Sunshine Coast Council's Kids in Action program (https://www.sunshinecoast.qld.gov.au/Environment/Education-Resources-and-Events/Kids-in-Action-Program/About-Kids-in-Action)

NZ – Since the Tree Tags contain the name of the specific tree species as well as specific measurements, is there a curriculum that was developed to facilitate the necessary assessment? How are the procedures for identifying the trees and taking the necessary measurements facilitated?

***JG* –** No curriculum was developed. All Tree Tags in schools have been supported by the local Council in which the school occurs. Through this collaboration, they have had me go in as a consultant and talk to the students about what we measure and why and then how we analyze the data – this presentation varies depending on the age of the students.

The trees to be measured have been pre-selected by Council and the school and we identify the tree species during this process. Then, depending on how old the students are, they gather some of the data (e.g. DBH, crown spreads) and I take measurements of the rest of the more complicated structural data (e.g. tree height, % crown missing, crown dieback). I then collate and model the data through Eco, convert the outputs to the relatable metrics and provide the schools and Council with the outputs they need for their Tree Tags

NZ – Given that the students are conveying tree information to the general public, are the students also being taught about trees – what makes a tree a tree, how they grow, what they need in order to grow well into maturity?

***JG* –** Yes, this sort of teaching opportunity is being run by the schools depending on their own curriculum. There are also teaching opportunities in math, English, art, and conservation science.

NZ – What sort of tools are the students using to do the height measurements? Are the schools helping students make their own clinometers or…? Did the students have any problems understanding how to take the measurements? Does the curriculum explain why these certain measurements are important – what they tell us?

***JG* –** Depending on the age of the students, I vary how and what measurements are taken. For example,

- the daycare center kids (ages 3–4) did no measurements (I did all of them) but the teacher talked with them about the importance of trees. As a result, the kids drew pictures and we used these pictures and some of their quotes on the back of the tree tags.
- the year 5 ESL class (ages 9–10) – we taught them how to measure DBH and canopy spread. With 3–4 students per group, DBH was measured using a measuring tape and then canopy spread measured by 2 students standing under the widest extent of canopy and another student or two using a rolling meter counter to measure the distance between the kids standing under the canopy extents.

- the year 6/7 students (ages 10–11) were taught how to collect all of the data, though I still collected all data as well to check accuracy. The height of the tree was estimated using a meter stick, applying the equivalent triangles concept. We love finding ways to teach kids "tree-gonometry"!

NZ – If an I-Tree Eco project is being established, what is happening with the resulting output beyond its use to create the Tree Tags?

***JG* –** The raw Eco data is provided to Council and is able to be collated in to their existing database to help gradually build the tree database information (most local Council's here do not even have a full tree inventory of location and species, let alone Eco information….this is gradually changing).

NZ – Have you thought of expanding the Tree Tag initiative to include other actionable endeavors such as teaching students how to care for trees that have been planted – how they can assist with growing their trees? This would address the issue of how we can enable trees' capacities to deliver those critically important Ecosystem Services benefits.

***JG* –** This is a great idea, but is outside the scope of work we generally do. If a school wanted to contract my time to provide some of this information we could do it, but usually schools are not in a position to engage a consultant. The collaboration with local Council though is a good resource for schools and some have had the Council arborist visit and talk to the kids about other tree-related issues.

NZ – Have you thought of actionable ways in which to further promote the Tree Tag initiative – t-shirts for the participating students announcing that they're a Tree Tagger and showing a really creative student-produced tag or creating an exhibit of the tags or, once you have more schools involved with a broader Australian representation, creating a contest for a school Tree Tag project.

***JG* –** These are all great ideas – and yes, we've thought of various ideas that create a type of competition and potentially gameifying things.

NZ – Have you initiated interest from local media regarding the Tree Tag data and what that means to the local community?

***JG* –** Media interest would be generated by the Council or School themselves, not by us as the consultant.

Washington D.C. – Casey Trees Urban Tree Inventory/i-Tree Eco Assessment

Casey Trees is Washington D.C.'s premier urban forestry not-for-profit. The organization, established in 2002, has been at the forefront of D.C.'s urban greening. Their mission – "…restoring, enhancing and protecting the tree canopy of the nation's capital. To fulfill this mission, we plant trees, engage thousands of volunteers of all ages in tree planting and care, provide year-round continuing education courses, monitor the city's tree canopy, develop interactive online tree tools and work with elected officials, developers, and community groups to protect and care for existing trees and to encourage the addition of new ones."

As with so many cities throughout the world, the District of Columbia has seen a steady decline in tree canopy cover. To address this issue, "Casey Trees has set a goal of attaining 40 percent canopy by 2032; planted over 25,000; educated thousands of residents about the importance of urban tree canopy; supported the tree planting efforts of the D.C. Government, the National Parks Service, community groups and residents alike; inventoried and tracked the District's tree resources to promote continued public funding for D.C.'s trees; advocated for green, tree friendly development and similar pursuits."

Washington D.C., with the involvement of Casey Trees, has been implementing i-Tree Eco assessments since 2006 https://caseytrees.org/wp-content/uploads/2014/04/report-2006-09-ufore2004.pdf and again in 2010 https://caseytrees.org/wp-content/uploads/2014/04/report-2010-01-ufore2009.pdf] and 2015 https://caseytreesdc.github.io/itreeeco/?_ga=2.185549024.856837298.1575888836-223817048.1573990171, with field data collected by an array of interns, citizen foresters and volunteers (sponsored by Casey Trees Endowment Fund) and by students from the University of Maryland Urban Forestry Program.

To facilitate citizen participation in i-Tree assessments, a citizen science module was created. Following are the questions posed regarding Casey Trees Citizen Science endeavor with answers provided by Casey Trees' Director of Science and Policy and ISA Certified Arborist, Jessica Sanders, PhD.:

> **NZ** – I can see that the Citizen Science web page begins with an outreach for Citizens to get involved with inventorying neighborhood parks and green spaces. How is an awareness of this activity disseminated? Is this outreach listed on other organization websites? Are there articles in local newspapers or in neighborhood weeklies? Do you use countertop displays to distribute brochures or other pertinent information?

JS – Our volunteer program at Casey Trees includes a host of things. We advertise to our house file of over 13,000 email addresses. We also target specific communities for our park inventory program and send out information to our locally elected Advisory Neighborhood Commissioners as well as an outreach to local civic/community/neighborhood associations. We frequently get groups that contact Casey Trees for volunteer opportunities. We have also advertised volunteer opportunities in the Washington Post as well as other local newspapers.

> **NZ** – For the original I-Tree assessment project, I saw that you used sample plots. Did those sample plots include street trees or other isolated plantings or were the plots only within parkland or other open landscape scenarios?

JS – For the i-Tree assessment project – there were 201 random permanent plots. The plots were placed randomly throughout the city – which means that they fell on streets, street trees and all other land in D.C.

NZ – I also saw that the training was on site. Who did the training and how did the trainers prepare? Were they professionals who had previously done an I-Tree Eco project or was this new to them as well? Did the trainers create handouts to facilitate an understanding of the measurements which would have to be taken and exactly how to take them?

JS **–** I conducted the training on-site. I prepared a deck of slides as well as an outdoor activity to ensure that the participants had an accurate knowledge of what was expected of them. We had a comprehensive training as well as a per-volunteer time slot that included at least 2 staff members (usually interns), hired and trained specifically for the i-Tree Eco assessment.

NZ – What was the age range of the volunteers? How many of the 50 were youth and how many were adults?

JS **–** The age range of the volunteers was from 6 months to 82. The volunteers contributing the most work hours were typically retired and could work during the day. We had several volunteers that committed to 8 h shifts on the weekend. Not all 201 plots were visited by volunteers – because the plots were placed across the city – we wanted to ensure the volunteers had a good experience – therefore any plots that were on asphalt with no or a very small amount of trees as well as 'tricky plots' (ex. Flamingo cage in the national zoo) were visited by staff only. Our interns/staff typically used bicycles for transportation and volunteers used public transit, walking, personal automobiles and bicycles to get to the locations.

NZ – Was there any preliminary dialogue about trees, why they are so important, what makes a tree able to deliver the ecosystem services benefits I-Tree Eco is looking to assess and how do the measurements provide that information?

JS **–** Yes – we had several slides that talked about the benefits of trees. After the assessment was complete, information was disseminated on why trees are important.

NZ – Once the volunteers were trained and the project began, did the trainers remain on site? Were there measurements that proved more difficult than others? Were the volunteers able to work with the clinometers without any difficulty? Were they able to manage the dieback condition assessment as well as crown light exposure?

JS **–** Yes – there were always staff on site which ensured the quality of the data. In addition, if someone got injured there was someone who was CPR/First Aid Certified. I would say that trained volunteers were less sure of themselves than paid staff, as such – there were several measurements that were harder to take. The volunteers had laser rangefinders – which we had already purchased for the park inventory program. Overall technology was a harder entry point for volunteers which is why it was helpful to have staff that was knowledgeable to help facilitate the measurements.

NZ – How have you kept the volunteers you have trained involved with their trees? Are they doing anything to improve their neighborhood trees' health by performing such things as simple management tasks, e.g. mulching bare earth or irrigating during drought?

JS **–** Yes – we keep in touch with the volunteers. Our volunteers prune, water, mulch and provide general tree care.

NZ – Is there specific training for such tasks as pruning, mulching, etc.? Can you share the training information or direct me to the online link?

JS **–** We have stewardship-level volunteers who water, weed, mulch and care for the trees post planting. How-to information can be found at the following link: https://caseytrees.org/take-action/water/ https://caseytrees.org/resources/?search=&resource_category=how-tos

NZ – I see that there is a watering protocol on the website. This prompts some questions. Has Casey Trees thought about funding for the 20-gallon TreeGator bags or an equivalent in your tree planting program which would greatly facilitate slow drip watering which would not necessarily require owning a hose?

JS **–** We currently use Ooze tubes on all trees we plant.

NZ – How has the resulting I-Tree data output been used? Is it incorporated into D.C.'s municipal management strategy? Has the data been added into D.C.'s official inventory?

JS **–** The i-Tree data output has been used in the Tree Report Card, in our own science communications as well as incorporated into D.C.'s management strategy. D.C. is a little different as it has a large number of Federal properties – several of those management units use the outputs of the i-Tree data in their management strategies as well.

NZ – The Citizen Science page also contains a "Nominate a Park" entry. Does your mobilizing involve already trained volunteers or is the nominated park's neighborhood tapped for volunteers or is acceptance of the nomination partially based on the neighborhood recruiting the volunteers? Does each accepted park go through an I-Tree Eco assessment?

JS **–** We pick parks for the park inventory program based on partnership priorities as well as the nomination process. We also use that nomination page to get people interested and excited about the program. We run each park through the i-Tree suite and then update all the benefits on the park inventory map.

NZ – Has there been any interest in expanding the neighborhood green space concept to include street trees in the assessment?

JS **–** Washington DC's street trees (not on federal land) are managed by the District Department of Transportation – Urban Forestry Division (DDOT-UFD). DDOT-UFD has a comprehensive inventory of the street trees in the city that is regularly updated and maintained.

NZ – Since citizen advocacy is so critical to how a City plans to accommodate its urban trees as well as fund its management operations, I was hoping you might share advocacy aspects of Casey Tree citizen activities, specifically the Advocacy training, e.g. relevant information such as specific training information, a summary of the curriculum or any other pertinent information?

JS – The advocacy program is under my direction within the Science and Policy department.
https://caseytrees.org/take-action/advocate/ We host a workshop that teaches people how to advocate for trees in their areas of the city.

Planning and Design

There is little in the architecture of a city
that is more beautifully designed than a tree.
Jaime Lerner,
Brazilian Architect & Urban Planner

Celje Slovenia – Urban Forest Planning with User Participation

Celje is the third largest city in Slovenia situated at the confluence of the Savinja, Hudinja, Ložnica, and Voglajna rivers in the lower Savinja Valley.

Viewing a satellite image of the City will show that Celje is surrounded by forested and agricultural lands but that it also has a City Forest which is part of its Urban Forest. Located along the southern edge of the City, this Town Forest, with its 114 hectares of forested land and its 15 km of well-maintained walkable trails, is the largest public green area within the City.

At the beginning of the 1990s, recognition of varied interests and potential for Celje's Urban Forest brought with it a new approach in Urban Forest planning and management. "In 1996, the municipality of Celje accepted the initiative of the Forest Service and started to implement the strategy of Urban Forest development which included the protection of the Urban Forest through local law, the improvement of the ownership structure with the redemption of private forests, the development of recreational and educational potential, the development of adequate public relations and assuring stable financial resources for the management. During the last two decades, the concept of Urban Forest governance based on close co-operation between the local unit of the Slovenia Forest Service and the municipality of Celje has evolved" (Hostnik, R. Urban Forests of Celje, Slovenia) which is evidenced by the diversity of partners, consisting of the Municipality of Celje, the Celje Unit of the Slovenian Forest Service, private forest owners, forestry contractors, recreational users of forests and local schools.

To expand on Urban Forest governance concepts and implementation of the user advocacy aspect, I interviewed my colleague Robert Hostnik, who graduated from the University of Ljubljana with a BSc in Forestry and an MSc in Nature Protection and has, over the past 27 years, integrated both Forest and Urban Forest management disciplines into his daily work including consulting and advising forest owners, production of forest management plans, participation in spatial planning processes, forest monitoring, silviculture, forest infrastructure development and maintenance. Although trained as a traditional Forester, Robert has embraced Urban Forest research and incorporated it into his work. "My main fields of interest and expertise are related to Urban Forestry, social forest functions, nature education and protected area management. For more than two decades, I have been actively involved in the development of Urban Forestry in Slovenia."

NZ – I know you are a trained Forester but how did you incorporate that into the Urban Forest concept/philosophy?

RH – The prevailing close-to-nature concept of forest management and so called *freestyle silviculture* is very close to Urban Forest management. It was a good base for my work. Of course, I had to add new knowledge from other professional fields, e.g. arboriculture (it was part of my master studies), governance, forest pedagogy, participation, public relations, conflict management. Over the years, we have also developed good cooperation with professionals from other fields, like landscape architecture, urbanism and pedagogy.

NZ – Although users of Celje's Urban Forest such as the City Forest, noted above, are not participating in any direct management activities, they are contributing planning information through a process of user surveys. Can you provide an example of the questions you ask users? Also, can you give examples of responses that have resulted in forest planning and management adaptations so we can connect the dots between citizen user input and planning/management outcomes?

RH – For example, we were asking regular Urban Forest users about their opinion of forest paths, maintenance, equipment, marking, forestry measures etc. Below is one of the examples (Fig. 5.5), where we were asking visitors, what is missing in the Urban Forest:

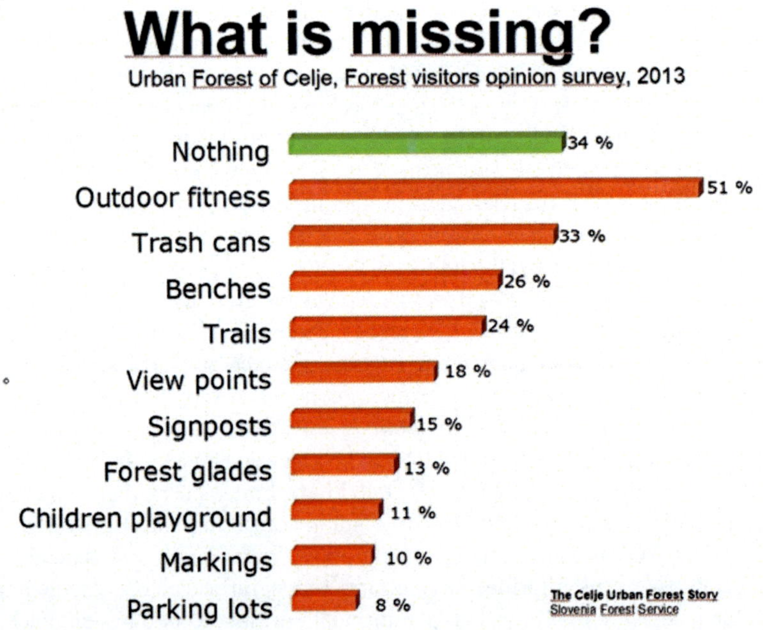

Fig. 5.5 Celje Urban Forest Forest Visitors Opinion Survey 2013. Robert Hostnik

We partly considered the result of inquiry in our next investment programme (eg. outdor fitness, benches, signposts, children's playground) but did not consider others that are in direct opposition with our management plan (eg. we don't want trash cans in the forest, or more forest paths, as we consider that the existing forest path network is large enough).

NZ – If there are planning meetings that affect the community's forest, do you include community leaders or reach out to known users to offer their input?

***RH* –** Yes, communication with community representatives is a constant.

NZ – Can you provide some information about this Urban Forest – how is it integrated into the community and what importance does it have for the community?

***RH* –** After 25 years of development and working for and with the users, it is very well integrated. It seems that a large percentage of the citizens have accepted the Urban Forest of Celje as their own forest. This statement is not based on any research or inquiry. It is my observation and feeling that comes from my communication with users on a daily basis. In recent years, we are also receiving important feedback through social media.

The Treehouse in the City Forest (Fig. 5.6) exemplifies how Robert and his team have invited users into Celje's Urban Forest and have, at the same time, facilitated their education, as stated on the travelcelje.com website (https://www.travelcelje.com/EN/a-town-forest-with-a-tree-house).

Fig. 5.6 Tree House. Urban Forest of Celje. Matjaž Čater

"The unique tree house, which is entirely made of Douglas fir (*Pseudotsuga menziesii*) wood from the City Forest, is designed as a small informative and educational center for the City Forest, intended for school groups and individual visitors to the forest. This is the largest tree house in Slovenia and is also well known abroad. From the initially modest tree observatory, it has grown into an open, universal meeting place, in which almost anything can happen: the house acts almost magically, from a magnet for individual, family or group and school visits, to becoming a real lab for exploring and eavesdropping on trees, a space for various creative workshops, outdoor and indoor concerts, poetry evenings and exhibitions ... The Treehouse is also subject to special expert attention from international foresters, biologists, ecologists, artists, and architects, who have already offered many professional recognitions."

In an article entitled Urban Forests of Celje, Slovenia, focused on Partnership Case Studies, Robert has offered the following list of the main lessons learned and conclusions:

- The existing governance covers the co-ordination between the Forest Service, the municipality and the local contractors as the main actors of Urban Forest management as well as the users' participation and private forest owners as the most important stakeholders.
- Strong, long-term oriented partnerships between actors produces synergy and cost effectiveness.
- The common thread of Celje Urban Forest management is characterized by high quality performance, cost effective approach and organic, gradual growth.
- The driving force with expertise, initiative, innovation and enthusiasm is the cornerstone for long-term success.

And, if success can be measured in the number of users of the Urban Forest resource, Celje's Urban Forest has seen more than a 3-fold increase, becoming "a role model for other Slovenian cities... The management plan from 2006 further developed the existing concepts and brought in new ideas. It was especially focused on (1) strengthening of governance and users' participation, (2) popularization and branding, (3) coordination of public and private interests (4) education based on forest pedagogy principles and (5) on further development of infrastructure and equipment for recreation, education and experiencing. The following years brought a stronger recognition and even wider popularity of the Urban Forest of Celje. Their users have become more active players in the decision-making processes... The Urban Forest has gained good political support and has become an integral part of the city's master plan... The future challenges for Urban Forest management include innovative marketing activities, further development of educational contents and implementation of instruments for balancing interests in private forests." (Hostnik, R. "Tailoring the Brand: Two Decades of Development of the Urban Forests of Celje")

São Paolo Brazil – The Atlantic Forest Municipal Plan

São Paolo is located in the Atlantic Coast restingas ecoregion within the Atlantic Forest biome, an area that stretches along Brazil's Atlantic Coast from Brazil's northeastern-most point at Rio Grande do Norte south to Rio Grande do Sul, extending inland into Paraguay and parts of Argentina. The Atlantic Forest is considered a biodiversity hotspot with only 28% of its native vegetation cover remaining (World Wildlife Fund: Atlantic Forest. World Wildlife Fund. 29 November 2011.) Although the region continues to support an extraordinarily diverse biome, the continued loss of forest combined with the ever increasing impact that humans have on the land and its available resources has prompted conservation activities in a number of areas and through a number of initiatives, such as the Atlantic Forest Law, established in 2006 https://www.ecolex.org/details/legislation/law-no-11428-on-the-use-and-protection-of-atlantic-forest-biome-lex-faoc069483/?q=Atlantic+Forest+Law&xdate_min=&xdate_max=

With the knowledgeable assistance of Dr. Maurício Lamano Ferreira, we will explore this law at its municipal level – Atlantic Forest Municipal Plan – and learn how it has been realized in the city of São Paolo with the participation of local residents. Mauricio, through his areas of study and his work clearly has his feet in both the forest world as well as the Urban Forest world. He is currently a professor and researcher in a Master's program at a university in São Paolo. In addition, Mauricio works as a volunteer consultant for the São Paolo State Secretariat for the Environment and thus is frequently involved in public projects. His undergraduate studies included Biological- as well as Geo-science and his Masters was in plant ecology, specifically the relationship between plants and urban air pollution. His PhD studies in the ecology of the Amazon and the Atlantic Forest focused on the pattern of tree growth and leaf litter production in parts of both the Amazon and the Atlantic Forest.

NZ – Please explain what the intent of the Atlantic Forest Law is?
MLF – The federal Atlantic Forest Law advocates the conservation, protection and regeneration of the Atlantic Forest, the second largest tropical forest in South America. However, due to the urbanization history of several Brazilian cities, this biome has only 7% of their forest remnants, making the law an important public policy instrument to restore the biome. "The aim is to achieve the sustainable development and the protection of biodiversity, human health, landscape, water resources and social stability. Particular attention is paid to the following principles: socio-environmental function of the property, inter-generational equity, prevention, precaution and polluter-pays…"

One instrument of this law to be implemented at the municipal level is the Atlantic Forest Municipal Plan, which involves, besides public actors, the population itself. In the city of São Paulo, this instrument was elaborated with popular participation, which participated in determining what would be the priority areas for conservation of green spaces within the urban region.

NZ – How has the Municipal Plan been mandated?
MLF – The Atlantic Forest Municipal Plan (PMMA) is integrated into the city's new Strategic Master Plan, sanctioned in 2014, and was developed by the City Hall, through the Municipal Green and Environment Secretariat (https://www.prefeitura.sp.gov.br/cidade/secretarias/meio_ambiente/).

Its preparation also includes the support of the Municipal Secretariats for Urban Development and the Coordination of Regional City Halls, the Fundação SOS Mata Atlântica, Sesc and the Universities.

The Plan determines priority actions and areas for conservation, inspection and recovery of native vegetation and biodiversity in the Atlantic Forest, based on the mapping of existing remnants in the city of São Paulo. The PMMA also encourages sustainable technological experiments, management of actions that reconcile the conservation of the biome with the economic and cultural development of the municipality, strengthening social organization and citizen participation in the management of public policies.

The preparation and execution of the PMMA can be carried out by the city halls and by the environmental councils of each municipality within the Atlantic Forest biome, and the municipal councils must be formed by a technical team and social groups. In São Paulo, NGOs, universities, representatives of commercial, industrial sectors and even union centers representing a specific class of worker form this council. The invitation to participate is open to everyone but the will to do this must come from the community. For example, in some strategic actions such as the creation of ecological corridors between natural parks, managers invite the community and people interested in attending assemblies or sending representatives.

It is important to highlight that most of the state capitals of Brazil are located in the Atlantic Forest biome, threatening natural conservation due to the expansion of cities. The purpose of the PMMA is to work at the municipal level in order to recover vegetation, whether through urban or rural reforestation.

NZ – How has the PMMA been realized in Sao Paolo, especially considering the continuing growth of the City?
MLF – In the case of the city of São Paulo, a special committee of the Municipal Council for the Environment and Sustainable Development was created to monitor the development of the work. In the Action Plan, six areas were created, and each of them has a link for popular participation.

The citizen could choose one of them or participate in as many as they wished, such as:

- Protected Areas and Green Area Systems;
- Inspection and Monitoring;
- Licensing, authorizations for vegetation management and environmental compensation;

- Environmental Education, Communication and Social Participation;
- Climate Risk Adaptation and Reduction
- Structural issues

For example, these are areas in which the municipality opens up for popular participation through park Management Councils. Management Councils are made up of neighbor park residents who represent a civil society in decision-making about park management. There are several areas of activity that the Management Council can act on, a decision which has more to do with its local reality.

The Management Council has an important vote in park management decisions, however, there are other collegiate bodies that also vote in these decisions, such as the government itself. The Management Council can give an opinion on actions to be carried out or prioritized in the management of the green area. The Management Council's participation is more in charge of the suggestion of guidelines. The management board rarely performs tasks, except at festivities.

NZ – What is the mechanism for citizens to get involved? Are there announcements that some aspect of the PMMA will be addressed at such and such a time…? Is there an online access to the list? Does someone do this directly through the Council?

MLF – Citizens as well as NGOs are invited by public call, usually posted on Internet pages (for example, the Municipal Green and Environment Secretariat website, etc.).

NZ – You have included environmental education in the participatory list

- what does this education consist of?
- how is it being offered /disseminated?
- have schoolchildren been included and, if so, is this environmental education tied to experiential learning?

MLF – Environmental education is the responsibility of specialized courses at an open and free university that provides free shortcuts to the population (details at https://www.prefeitura.sp.gov.br/cidade/secretarias/meio_ambiente/umapaz/

In addition, the Municipal Green and Environment Secretariat has an open university, that is, it has a building in Ibirapuera Park that promotes courses in gardening, botany, conservation, environmental education and other subjects, all free of charge for the population. These are short courses taught by specialists in the fields. https://www.prefeitura.sp.gov.br/cidade/secretarias/meio_ambiente/umapaz/

NZ – You also mention Communication and Social Participation.

- Is this through social media or…?
- Do the Councils have meetings that are open to the public?
- Is the information shared at the lay person's level of understanding?

MLF – Varied social media are also used to publicize public events. The Councils are essentially formed by the residents of the park however, it is a totally public and open assembly. In fact, information that is generated at meetings is recorded in the public minutes so that any lay person can have access.

It should be noted that the participation of the management board may or may not be associated with the PMMA's proposals, since they have a much broader function and discuss other matters besides PMMA.

NZ – Are there any examples of citizen participation which you feel were very successful? Can you please share them?

MLF – PMMA itself had a satisfactory popular participation in my understanding. The process was very democratic and disseminated in the traditional media of the federation for this type of call.

One of the most recent examples was about a decision to privatize Ibirapuera Park, the most famous and important park in São Paolo. The municipality has a lot of financial expenses to maintain this park, which is one of the main postcards of São Paulo. Therefore, the city opened public hearings to discuss with society the Master Plan that addressed the issue of the concession of the park. The following was published on the City Hall website "Valuable contributions were made at the second and last Public Hearing organized by the City Hall, through the Secretariat of Green and Environment (SVMA), for the democratic construction of the Ibirapuera Park Master Plan. Some contributions were made through letters and another 37 participants who asked for a word to defend the transparency and clarity of the process. The objective was very specific: the users and members of the Park's Management Council want the future Master Plan to specify clearly how it will guarantee the preservation of fauna and flora, as well as the exhibits used in the park.

""We understand that this democratic construction presupposes listening, taking notes and analyzing all the interventions made, and the result was extremely positive for Ibirapuera", evaluated the Coordinator of Parks and Biodiversity Management (CGPABI) of SVMA, architect Tamires Carla de Oliveira."

https://parqueibirapuera.org/?gclid=CjwKCAiAj-_xBRBjEiwAmRbqYjKrzLoIMVFA0zqS_uW3uItyT86dijM94WSJlXeOuTu-PLe5LYZg-RoCd2AAv

NZ – Does São Paolo have an indigenous population? If so, does that population have ownership rights to any of the Atlantic Forest lands within the São Paolo jurisdiction and how is their traditional connection to the land and their sustainable use of it being recognized and accommodated?

MLF – São Paulo has some Guarani indigenous communities. They have a demarcated area, although it is much less than was requested by indigenous leaders. The state subsidizes food and basic health services for indigenous people. In 2009, I worked on a project that sought to integrate the young Guarani into the discussion of public policies in the city of São Paulo, with the aim of directly involving the protagonists and not only assisting in the elaboration of policies. This project was based on Chapter 26 of AGENDA 21.

Winnipeg Canada – SpeakUp Winnipeg

In 2009 and 2010, citizens of Winnipeg came together and created Our Winnipeg, a 25-year development plan for everything the City does, providing "a vision and policies that influence the delivery of City services, how we get around the city, and decisions about how the city grows." https://speakupwinnipeg.ca

To facilitate citizen involvement in support of sustainable action, the speakupwinnipeg.ca website was developed to make participation in the entire planning and development processes as easy as possible. Documents supporting varied aspects of planning history and strategies were made available online https://speakupwinnipeg.ca/documents/ including a Call to Action, encouraging all citizens to take part, speak up and share their thoughts. The Call to Action initiative was realized via online surveys, written submissions, face-to-face community meetings and interactions at festivals and events.

One of the pillars explored through the Call to Action and the creation of a sustainable Winnipeg https://www.winnipeg.ca/interhom/CityHall/OurWinnipeg/pdf/CallToAction.pdf is the environment and environmental issues. In addition to a community greenhouse gas reduction strategy, the following Urban Forest ecosystem initiative was framed:

"*Respect and value our natural environment*:

- Take immediate action on climate change, both as a civic government and as a community
- Plant more trees and actively preserve our Urban Forest
- Increase opportunities to reduce waste and to compost
- Recognize and preserve parks, green spaces (including the Urban Forest) and riverbanks as green oases in our urban setting."

This initiative has seen the following initial actions set forward:

- *Greenspace Naturalization Standards* Establish public greenspace naturalization standards that would improve biodiversity, reduce maintenance costs and contribute to a healthier environment. This action will also include a public awareness campaign;
- *New Parks Strategy* Involve the community in a strategic plan for the sustainable development and management of Winnipeg's parks, public places and open spaces;
- *Explore Urban Agriculture* Through OurWinnipeg, and as a partner in a community-led initiative, explore opportunities to develop urban agriculture and local food security programs such as community gardening, urban farming and farmers' markets;

and the establishment of Direction #9 within the Sustainable Winnipeg concept: https://www.winnipeg.ca/interhom/CityHall/OurWinnipeg/pdf/ASustainableWinnipeg.pdf

Protect and enhance the Urban Forest
Enabling strategies:

- Increase the stock of trees through tree planting programs, and encourage tree planting by others;
- Adopt high standards of tree maintenance, replacement and protection during construction, and require developers to retain existing trees in new developments, wherever possible.
- Replace trees affected by disease, Dutch Elm Disease in particular, and diversify the variety of new trees;
- Encourage the participation of other levels of government in programs to protect and enhance the city's Urban Forest.

Tree Planting and Establishment

Plant one tree.
Plant it for a lifetime – its lifetime, not yours.
Plant it so that it may grow;
so that it may become the tree it has the capacity to become;
so that your children and your children's children
may revel in its presence.
Naomi Zürcher

Tree planting is the most broadly recognized and accepted initiative within the scope of participatory Urban and Community Forestry actions. Tree planting is *sexy*. It appeals to many human interests from celebrating an event or memorializing a loss to providing an opportunity to connect with community members, participating in a family-oriented event or just being outdoors. It is a highly visible action that businesses and politicians can get behind, fund and take credit for. A real win-win for many actors.

Given the extraordinary number of such programs, I tried to find models that spoke to more than just planting but rather had elements of planning or advocacy but also an underlying social ecology embedded in the initiative as well.

In setting the stage for the urban tree planting sampling, I can think of no better starting point than Joseph Beuys *7000 Eichen – Stadtverwaldung statt Stadtverwaltung* [*7000 Oaks – Urban Forests instead of City Administration*]. Joseph Beuys (1921–1986), the man in the felt hat, was a profound figure in post World War II contemporary art. He explored a vast array of art media – painting, sculpture, drawing, performance – from a very unique and often confounding perspective, investing symbols of civic, social, environmental and political commentary into his work, as his philosophy "art is life" evolved.

In 1982, Beuys used the provocative to instigate citizen participation in a profound example of social art action, envisioned through Urban Forestation – the planting of 7000 oaks in Kassel Germany "...to reanimate the lifelines of nature which are endangered by the inhumanity of our economy."

7000 Eichen was a product of a developing concept "When I thought of a sculptural form which could comprehend both physical and spiritual material, I was absolutely driven by the idea of Social Sculpture. …from now on my notion of art is a social one which means that we are not talking about a traditional understanding of art but we are talking about the possibility of every single human being in the social notion to be a participant in the social sculpture. And, only under this condition it must be understood when I say that every human is an artist. I am talking about a new discipline of art, the social art, which creates interpersonal relationships and which unleashes the individual potential of the free individual in a continuous and affectionate clash of individual ideas."

In a conversation with Richard Demarco (Scottish artist and gallerist), in anticipation of the first planting for documenta 7 in 1982, Beuys was asked why oak trees. He replied:

"I think the tree is an element of regeneration which in itself is a concept of time. The oak is especially so because it is a slowly growing tree with a very solid heartwood. It has always been a form of sculpture; a symbol for this planet ever since the Druids, who are called after the oak. Druid means oak. They used their Druids to define their holy places. I can see such a use for the future. The tree planting enterprise provides a straightforward but radical possibility for this when we start with the seven thousand oaks.

The planting of seven thousand oak trees is only a symbolic beginning. Contrary to its initiative, progressive features such as a symbolic beginning require a marker, in this instance a basalt column. Future goals for the project include:

- An ongoing scheme of tree planting to be extended throughout the world. This was envisioned as a part of a global mission to foster environmental and social change.
- Growth of awareness within the urban environment of human dependence on the more extensive ecosystem educational outreach.
- An ongoing process whereby the society would be activated by means of human creative will social sculpture."

In implementing the *7000 Eichen* project, Beuys was quoted as saying "The trees already recognize more of the hardship and misery of the people than the people do themselves". For Beuys, trees were like organs of perception… "as a principle of growth, of movement. *7000 Eichen* was a metaphor for a new commerce in which humanity drove creativity and its resulting products." https://publicdelivery.org/joseph-beuys-7000-oaks/

The goals that Beuys put forward have continued to live on in projects that *7000 Eichen* inspired including such endeavors as *Oaks of North Lawndale*, featured in the following sampling.

Beirut Lebanon – Regenerate Lebanon

I learned of this wonderful project as a result of an article entitled "How Lebanese environmentalists helped spark a green revolution" in the 5 January 2020 online issue of The Grist https://grist.org/article/how-lebanese-environmentalists-helped-spark-a-green-revolution/. (The Grist is "an independent, irreverent news outlet and network of innovators working toward a planet that doesn't burn…") https://grist.org/about/

Although the article focused on students and their initiative to pick up the trash that had accumulated from protests in Martyr's Square in downtown Beirut, it provided some background information on the chain of events which would inspire Lebanon's own green revolution. In October of 2019, a wildfire started in a forested area south of Beirut. Hot windy conditions combined with the lack of government intervention resulted in the fire spreading quickly, engulfing and burning more than 3000 acres.

Government's lack of attention to the likes of a wildfire of such magnitude often has the result of igniting an already over-taxed public and Beirut was no different as over 1,000,000 protesters took to the streets, demanding that the Government step down. By the end of October, the Prime Minister had resigned. Countries such as Lebanon often find themselves in the most bleak news of the day, reporting on varied problems and military actions. We generally accept that this is all there is when in fact, conflict often inspires positive action. Not only did the wildfires instigate the dismissal of an ineffective and unresponsive Government, they also sparked a green revolution, spearheaded by Adib Dada, an environmental architect and biomimicry specialist and Joslin Kehdy, who founded the multi-faceted NGO Regenerate Lebanon https://www.regeneratelebanon.com/initiatives/plant-lebanese

Fig. 5.7 Beirut Lebanon – Regenerate Lebanon/Beirut RiverLESS tree planting initiative. Adib Dada

Initiatives were already happening prior to the onset of the wildfire. In May of 2019, Adib Dada, through his company theOtherDada and in partnership with the social enterprise Afforestt, held an Urban Afforestation Workshop https://vimeo.com/348381254 (https://vimeo.com/377505669) (https://vimeo.com/385972577), as a result of which, over 1200 native trees and shrubs were planted in Beirut, using the Miyawaki afforestation technique (Fig. 5.7). https://www.afforestt.com/methodology

Adib has kindly agreed to share information about theOtherDada and Regenerate Lebanon's diverse greening actions, as follows:

NZ – Since you are a private company, how do you approach all the varied aspects of separating for-profit from social endeavor?

***AD* –** We are a for-profit company with a social and environmental impact. We identify as a Social Business, and are applying to become a Certified B Corp., as we believe businesses have the potential to have the most positive impact and can be a force for good. While we have revenue-generating private architecture and design projects, we've also been working since 2013 on Beirut RiverLESS, which is a self-initiated *pro bono* project focused on reversing the damaged condition of the city's most polluted river. We have also worked on many other *pro bono* projects. Given the current economic situation which has been declining over the last few years in Lebanon, the company is not making any profit, but is still committed to its environmental and social work regardless. We are looking into reaching a state of balance between our for-profit and non-profit work.

NZ – Are social initiatives part of the company's mission? Do you require that your employees /associates be like-minded?

***AD* –** Yes our team members have like-minded values of being driven towards working on projects with a social/ecological impact. The interview process is based less on specific skillsets, and more on what are the values of a potential team-member, and what new and different skills they can bring to the table. This goes hand-in-hand with our horizontal structure, where we believe each one, whether an intern or senior member, has something to teach the rest, hence roles are assigned depending on skills instead of seniority, and projects are shaped depending on skills available at the time.

NZ – My own story has possible similarities. I too had a private consultancy to deal with my work as a Consulting Arborist – for which I was always paid. All the work I have done and continue to do as an Urban Forester is *gratis*. The Urban Forest belongs to everyone and therefore it also belongs to me. I feel that whatever I do for the Urban Forest to improve its health and well being also improves my health along with all other residents, be they human or otherwise. How could I charge for that? Would be nice to hear a brief version of your story which I like to include in presenting the interviewee so the reader has a sense of who is behind the program.

***AD* –** Professionally, I am an architect, having earned my BAr in Architecture at the American University in Beirut and a Masters in Interactive Telecommunications from NYU. I also completed the Biomimicry graduate certificate program from the

Biomimicry Institute and Arizona State University. Based on Biomimicry, my work through theOtherDada promotes a symbiotic relationship between nature and the built environment by exploring new ways of creating generous and regenerative buildings; in essence, developing creative solutions that have a positive impact (Adib has been recognized in Apollo Magazine's 40 Under 40 as a Patron of the Arts in the Middle East.)

In regards to our urban afforestation projects, the pilot project was completely *pro bono*, our company even contributed 20% of the total funding for it. Our projects are always targeted towards communities, reclaiming public spaces and creating healthy spaces for communities to gather and for wildlife to return. However, in an effort to continue to make forests, we are working on becoming financially sustainable and making enough profit from private afforestation projects to cover our office and team costs for public afforestation projects. Our Beirut RiverLESS initiative has been running for the past 7 years without any funding.

NZ – Is your company a part of the Regenerate Lebanon initiative or are you part of that as an individual? How are responsibilities divided up between the two entities? Are you the initiator and invite them to join or vice versa?

***AD* –** I am a founding member of Regenerate Lebanon (RL) as an individual. However, the afforestation initiative serves the same goals as RL, therefore my team periodically collaborates with RL for specific projects related to reclaiming public spaces, creating forests for health and food, sustainable urban farming and consulting on regenerative farms. At RL, we are working on making self-sufficient neighborhoods, to short-circuit the linear funding model and create a more circular model where recipients are not only reliant on NGOs. Hence, our neighborhood interventions typically include a kitchen, community center, urban farm and rooftop farming, chicken coops and sometimes goats, tiny Urban Forests, as well as renewable energy and rainwater collection, waste management…

NZ – In the press release you had provided, there are five Miyawaki forest initiatives. How did you recruit participants for each and what was the most effective?

***AD* –** While we are only two people from the team working full-time on the afforestation initiative, we've had multiple ways of recruiting people for each of the forests. During the pilot project in May 2019, we collaborated with Afforestt to give a 6-day workshop and teach 15 participants how to create forests using the Miyawaki technique. The workshop participants were therefore working on creating the first forest from a mature forest survey for species identification and selection, to excavation, planting and maintenance. The next project in the cemetery in Beirut, was also completed with the help of some of these participants since it was finished directly after the pilot project. During the second phase of the Beirut RiverLESS forest project in November 2019, we recruited a paid intern to work on the logistics of the forest-making process, in addition to having 2 committed volunteers to help with the 3 week-long process.

Additionally, we had an influx of volunteers that helped us with trash removal, compost-tea making, mound-making, and planting the 1200 saplings. All the recruitment was done over social media, using Instagram and Facebook as the main channels to reach out to the community. For the downtown guerilla afforestation action, which we initiated as a form of protest during the October 17 Revolution, we had 10 individuals helping us with the process – from our trusted volunteers to revolutionaries sleeping in the tents on Martyrs' square, as well as police and army men who were stationed there to quell the protests! A very interesting/conflicting mix of people coming together for positive action. Since this project was tightly related to the October 17 Revolution of Lebanon, the 10 volunteers were easy to reach out to as they were interested in environmental activism and wanted to participate in direct action to afforest the leftover planters found along the protest area. In regards to the private Urban Forest (T-Forest), the main participants were the clients: the family residing in the building who owns the plot, as well as all the guards, workers and drivers from neighboring buildings.

NZ – Are the predominant participants youth/students? Are there older people participating? What would you say is the age range and is everyone from Beirut?

AD – The predominant participants consist of students, recent graduates, and working professionals related to the field of environment/architecture/design. The age range would be 18–40 and almost everyone is either residing in Beirut or visiting Beirut from abroad (Lebanese diaspora, foreign students, international researchers, journalists, artists…) I believe our youngest volunteer [my daughter] was 1½ years old, and the eldest volunteer in her sixties.

NZ – When you did the 1st and 4th Forest plantings in Sin el Fil, were participants from Beirut or from Sin el Fil?

AD – Most of the participants were from Greater Beirut, however we had many individuals and families from Sin el Fil participate in the planting as well as in subsequent maintenance and compost-tea making sessions.

NZ – I read that most of your funding for the varied initiatives comes from SUGI crowdfunding. I am assuming the primary costs are plant material, soil organic matter, spades, stakes. A big ticket would be renting an excavator along with the operator. If you had to equate the cost per seedling planted, what would that average be?

AD – The average for seedling planted would be $16 USD, which is less costly than the average $20–25 USD which other reforestation initiatives in Lebanon provide. However, this cost does not include machinery and watering, which was provided by the municipality.

NZ – How was the participation of Afforestt funded – were they donated services or did this also come from funds raised through crowdfunding?

AD – The participation of Afforestt was privately funded from corporations and individuals, in addition to my personal contribution.

NZ – Who is providing the seedlings and the soil? Does the fact that this is a social initiative help deflate the costs? Have you tried to get any of the materials donated and what has the response been?

AD **–** The seedlings and biomass are all purchased from local nurseries and initiatives that propagate their own native seedlings and create their own local organic biomass. Some minor materials were donated such as tools, shovels, buckets, however the bulk of the materials were purchased. We believe in empowering and providing economic opportunities to these suppliers, especially the native nurseries.

NZ – Who is able to attend the Urban Afforestation Workshop – project partners or volunteers as well? If only partners, how is the information critical to creating the forests passed along to volunteers? Do you have brochures or is there information posted online?

AD **–** The workshop's attendees were selected based on background and interest via an online application form. The attendees we selected were: artists, educators, students, scouts, and members of the local community. The information for creating forests was covered in the intensive and very detailed 6-day theoretical and hands-on workshop for the 15 participants and our team. General information is passed on to the volunteers via volunteering sessions where one of our team members at theOtherDada gives a thorough explanation of the session tasks, enabling participants to perform these tasks with full knowledge of the process.

Our project information is published online, http://sardassets.lau.edu.lb/arc_catalogs/the-place-that-remains-proceedings-essay-6.pdf and a brief project information leaflet about Beirut RiverLESS was produced as a handout.

NZ – Besides an introduction to the Miyawaki Technique, does the workshop offer anything that teaches about the forest as an ecosystem? How forests evolve and grow? How species within the forest interact and form associations?

AD **–** Yes, the workshop offered the participants a thorough detailed guide on the concept of PNV [potential natural vegetation] and Dr. Miyawaki's research and findings. The workshop covered the process of creating a Miyawaki forest from plant selection and botanical survey, excavation and biomass mixing to compost tea-making and planting, staking, rope-tying and mulching.

NZ – Are planting techniques described? Is the probability of compaction from the use of the excavator discussed and remedies offered?

AD **–** Yes, the probability of compaction is covered during the implementation of the forest. During the mound-making process the participants were observing the method used to level the mixed soil, where the machinery is instructed to fill the mounds and move backwards without moving over the prepared soil in order to avoid compaction. In addition to that, the soil is mixed with a huge amount of perforation biomass, which mitigates the compaction problem.

NZ – Is there a recommended mixture of soil to organic matter and, if so, what is that proportional relationship?

AD **–** Yes there is a recommended mix of soil to organic matter depending on the texture and chemical composition of the existing soil. The soil's texture is studied and determined on site by hand, and the carbon and nitrogen and organic matter found in the soil is tested in a lab. The soil is then mixed with certain amounts of perforation material [shredded straw], soil nourishment [organic compost, mushroom compost] and water retention material [husk] according to a table guideline depending on the soil's existing state. The ideal state of the mixed soil should be one that allows water to infiltrate, allows plant roots to grow in ease, and contains the right amount of organic matter to nourish the plants. Last but not least, the saplings are dipped in the compost-tea before planting, and the whole site is irrigated afterwards with compost tea to bring the soil back to life and support the saplings. Compost-tea is prepared and used every 3 months over the 2–3 year maintenance period as part of the irrigation strategy. After that, the forest becomes completely self-sufficient.

Chicago Illinois USA – Oaks of North Lawndale Tree Planting Program

The School of the Art Institute of Chicago (SAIC) was founded in 1866 as the Chicago Academy of Design. In addition to a "broad and dynamic spectrum of study in art and design, SAIC "has a number of institutional initiatives that provide the resources and support needed to create awareness and encourage participation in programs that address environmental challenges, diversity, research and collaboration, community partnerships, and the sharing of knowledge between disciplines." (https://www.saic.edu/about)

In that spirit, the project *Oaks of North Lawndale* offered SAIC a collaborative opportunity, together with the City of Chicago and the community of North Lawndale, "to reimagine the neighborhood as a verdant, peaceful and tree-lined place."

I first learned of this project during an online i-Tree webinar in March of 2018. The presenter, Taykhoom Biviji, offered information on SAIC's approach to the revitalization of a disenfranchised community through a social-minded greening endeavor. Originally from Mumbai, India, Taykhoom Biviji received a Masters in Arts Administration and Policy (2017) from the School of the Art Institute of Chicago (SAIC) and holds a Masters in Ancient Indian History, Culture & Archaeology (2015) from St. Xavier's University-Mumbai.

He has worked in the North Lawndale community for the last 4 years, first as a researcher prototyping incubator models for not for profits and then as the Homan Square Community Coordinator for SAIC. Currently, he is a Project Manager at the SAIC's Office of Engagement focusing on *Oaks of North Lawndale* and developing partnerships and engagement programs.

Taykhhom has kindly offered to share the *Oaks of North Lawndale* program with us, as follows:

NZ – What inspired SAIC and their vision for the project's initial concept?
***TB* –** SAIC was invited into the neighborhood in 2013 in response to the communities need for an art and design center. SAIC saw this as a manifestation of its core

value of 'Chicago'. Through community consultation in 2016 SAIC at Homan Square was launched. Through our conversations with the community our understanding of the need for trees, one particular story stuck out. Residents explained that when they drove to neighboring Oak Park, they would recognize it immediately because of the tree canopy coming into view and they would roll down their windows to let in the fresh air. We wondered why this was not possible in North Lawndale, leading to the discovery that the neighborhood has one of the lowest tree canopies in Chicago at 16.66%. We took inspiration from Joseph Beuys' *7000 Eichen* and in collaboration with the North Lawndale community, developed *Oaks of North Lawndale*.

NZ – How did Joseph Beuys *7000 Eichen* inspire *Oaks…*? Beuys referred to the project as Social Sculpture and involved the community of Kassel Germany in the plantings. How do you see Oaks… vision dovetailing with Beuys' social initiative?

***TB* –** We're not the first in the neighborhood to advocate for trees. Block clubs, church groups high schools and North Lawndale Greening Committee as well as the North Lawndale Community Coordinating Council (NLCCC) were already working towards planting trees. We see *Oaks of North Lawndale* as a response to the community's needs. We support their goals for a peaceful and verdant neighborhood through the tree plantings and all the programming around it.

NZ – I know that Pedro Reyes began his Palas por Pistolas project in Mexico in the city of Culiacán. How did Pedro Reyes get involved with *Oaks…*? Did he recruit local blacksmiths and carpenters? Was their involvement donated services or were they employed through the project?

***TB* –** Prior to starting *7000 Eichen*, Beuys did a Palas-style project where he created shovels based on the British world war 2 shovel prototype. We also were particularly inspired by Beuys' words "you don't make war when you plant trees". As we were creating the *Oaks* project for North Lawndale, we were conscious of the challenge of gun violence faced by the community. Pedro's *Palas Por Pistolas* felt like a natural fit and the right way to launch this project. We consulted with the community and they agreed. The forge that made the shovels is a local four generation forge within the community. They were compensated for their time and effort. The shovel handles were made by a local west side organization Greater West Town Training who have woodworking programs for neighborhood youth. A shovel was presented to them for their effort. The wood for the handles was acquired from the Park District that had cut down ash trees due to the emerald ash borer beetle – EAB. Further, the shovel handles were coated with bees wax purchased from local bee keepers.

"Palas por Pistolas" was initially conceptualized as a response to the gun violence in the western Mexican city of Culiacán. Reyes organized a campaign for residents to turn in their weapons to be melted into shovels that were then used to plant trees. During the campaign, 1527 weapons were collected and taken to a military zone to be crushed by a steamroller before shovels were created out of the weapon waste. http://pedroreyes.net/palasporpistolas.php?szLang=en&Area=work

"Ten years ago, I started this project, Palas por Pistolas, to withdraw weapons from circulation in my home country, and 10 years after, I am grateful to have the opportunity to do a new installment of this piece in the city of Chicago," said Pedro Reyes. "Guns spread death and suffering in the world, so I believe in making an effort to reduce the number of weapons out there, even if it seems an insurmountable task." About the planting of trees, Reyes adds: "We are starting to see the catastrophic effects of climate change, so another crucial undertaking is to constantly plant trees; our survival depends on this. This is the transformative essence of this project: to turn an instrument of death and turn it into an instrument of life."

The symbolic act of showing how an agent of death can become an agent of life will be reproduced during the *Oaks of North Lawndale* event on Saturday, September 23 from 10 a.m.–4 p.m. at Nichols Tower, 906 S. Homan Ave." The shovels are Pedro Reyes' works of art and are being sold to support the project. https://www.saic.edu/press/artist-pedro-reyes-reproduce-his-work-"palas-por-pistolas"-guns-shovels-north-lawndale

NZ – The video shows students producing drawings, etc. Was this part of the Guns for Shovels component? How did the students become involved and how were the resulting works of art incorporated into the project?

***TB* –** The kick off for *Oaks of North Lawndale* was a community event that saw around 500 residents attend. We wanted to create opportunities for residents to engage with the process of the melting guns. In partnership with our Art Therapy department, we developed a program that allowed those who have experienced trauma due to gun violence to write messages, make drawings and use them to wrap around the gun parts which were then melted. Additionally, the melting of weapons waste also created more material than what was needed to produce the shovels since you have to keep the foundry running. Students were introduced to the process of sand casting where the excess metal was poured on their etchings on the sand blocks and then given to them to take away.

NZ – Especially in disenfranchised communities, there is great value in tapping the local community for a training program. Not only does it boost the local economy and create a pathway to professions which otherwise would not be considered, it makes it easier to connect local residents to the people who are professionally providing care for the resource. From this perspective, how has the lay community become involved?

***TB* –** *Oaks of North Lawndale* was developed keeping in mind the high unemployment rate within the community and the barriers that some residents face in attaining gainful employment. One of our partners is the North Lawndale Employment Network (NLEN) with whom we're working towards developing a workforce employment program that trains the next generation of trained individuals for landscape construction, urban forestry and urban agriculture industries. As part of the kick-off, we were able to work with 10 NLEN clients and share with them the skills needed to plant and care for trees. The specific training for these clients was developed by our community partner, North Lawndale Greening Committee in collaboration with the local nursery Homan Grown. For the past 2 years, we have been working on pilot projects to better understand the sustainability of the project and how best to respond to the needs from the community. This has taken on two forms, one includes developing a Food Forest using the principles of Permaculture with the high school CCA Academy and the other is partnerships with local stewards.

CCA Academy provides non-traditional students, ages 17–21, who have completed eighth grade with a second chance to earn a high school diploma. Urban Agriculture is at the heart of their curriculum. The academy owns 6 contiguous vacant lots adjacent to their campus and was looking to transform them into a classroom focused on Urban Agriculture. In order to achieve this, SAIC students from the Architecture, Interior Architecture and Designed Objects (AIADO) departments, led by faculty member Odile Compagnon, partnered with CCA Academy high school students to develop designs for 7 different food forests over a semester. Perma Park was the design that CCA chose after a design charrette. We have planted over 75 trees on these 6 lots and they are being cared for by the high-school students as part of their experiential training. This also responds to the neighborhood being a food desert and the food forest being a source for fresh food within the community.

To fund raise for this project, our graduate students in the MA Arts Administration and Policy program worked with staff and faculty at CCA Academy to develop mission and vision statements for the food forest.

Another form of partnership has been with local parks and block clubs. One such partnership is with the residents of K-Town in North Lawndale. Located at 1900 S Kostner Avenue, Unity Park a community managed park on Neighborspace land. The community had been looking for ways to revitalize the park and bring renewed energy to it. To date, we have hosted two planting days at this site where residents learned about tree care and maintenance while planting a total of 40 trees. The second planting day also saw the acclaimed musician Yo Yo Ma partnering with the community to plant trees.

NZ – Are any of the simpler maintenance tasks, e.g. mulching and irrigation, being shared with the interested community?

TB – Yes, at all our planting events we partner with local nursery Homan Grown and the Chicago Region Tree Initiative (CRTI) to share best practices for tree planting and care. The tree planting and after care practices and procedures that CRTI developed are shared online at http://chicagorti.org/programs/treecare

NZ – If community members have an interest in lending a hand, is there an educational component that provides information on tree basics?

TB – One of our planting days was in partnership with an elementary school Chicago West Side Christian School. Here there was a specific interest in the biology of trees and the faculty and students developed informational panels about benefits of trees which included how much oxygen a tree produces and carbon it sequesters.

NZ – Did Citizens – youth or adults – take part in the I-Tree Eco assessment and, if so, what were their assigned responsibilities and what sort of training did they receive in order to execute those tasks with some accuracy?

TB – During the assessment, the community participated in answering a few questions but overall their involvement was minimal.

NZ – Has the I-Tree assessment data output been given to the municipal agency responsible for management of the Urban Forest resource and, if so, how has it influenced their management of the resource; if not, is there a plan to make the resulting data "actionable" outside the project?

TB – We have shared the study with the City of Chicago and the Chicago Park Districts. The then Mayor Rahm Emmanuel indicated that he would work towards making North Lawndale and Chicago greener.

NZ – Are there any planned activities, e.g. community tree planting days or community mulching events which connect the trained local workforce with interested citizens?

TB – We host such events every planting season. We generally have success with them in Fall but we have had limited success with the weather and survival of trees in Spring. At all our tree planting days, we train local residents in best practices for tree care and maintenance.

NZ – Has the project inspired incorporation into local school curriculum which would begin the process of engaging people and nature at a much earlier age and, additionally, bring a better awareness of all the great green careers that could be pursued which would contribute to the community's well-being?

TB – Currently, the engagement for local campuses is to increase the tree canopy around their campus. An ambition of the project is to work with Gardeners, a local not-for-profit to develop curriculum around nature and tree maintenance.

NZ – All too often, we see disenfranchised communities with a disproportionate percentage of criminal offenders. Has the project tapped into that part of the local population for the training program? if so, has there been any interest in creating a prison outreach program to facilitate transitioning ex-offenders from prison to community by engaging them in the project?

TB – Through our partnership with North Lawndale Employment Network (NLEN), we have been working towards developing a training program geared towards recently paroled citizens. This is currently something we're working towards.

Edmonton Canada – Ribbon of Green + Root for Trees Tree Planting Program

The City of Edmonton has involved its citizenry in city planning and resulting initiatives in every aspect of city life. This is no different for the planning and managing of Edmonton's Urban Forest as is evidenced by its 2012 Urban Forest Management Plan (UFMP).

https://www.edmonton.ca/city_government/documents/PDF/Urban_Forest_Management_Plan.pdf. This 10-year management strategy "…provides strategic direction for Edmonton's entire Urban Forest. This includes all trees within city limits – whether planted, naturally occurring or accidentally seeded. Trees found in parks, natural areas, the river valley, ravines, roadways, roof top gardens, commercial, residential and private lands are all part of the Urban Forest, and within the scope of this plan." The City has acknowledged, through varied studies and an i-Tree assessment that "Edmonton's trees represent an irreplaceable asset" and "Unlike other municipal infrastructure, trees increase in value over time… Edmonton's Urban Forest is an important resource that provides direct and tangible environmental, ecological, economic and social benefits to the city and surrounding communities. This valuable ecological network improves our air quality, reduces energy consumption, keeps soil from eroding and conserves water resources. With careful stewardship, these benefits can continue for generations, appreciating over time."

Leading up to the formulation of the UFMP, Citizen Surveys were conducted in 2009 and 2010. These surveys' results – "What We Heard" – indicated strong community support for the Urban Forest, the proposed Plan to establish its care, protection and preservation and the strategies being proposed to accomplish these tasks as well as being dissatisfied with the amount of care and maintenance information being shared with the Public. The surveys also provided Citizens with what each taxpayer was being asked to contribute to care and maintenance with almost all responders saying they approved or wanted the amount to be higher. https://www.edmonton.ca/residential_neighbourhoods/PDF/UrbanForestry_WWH_February2020.pdf

Throughout the UFMP, there are sections that specifically refer to the inclusion of the Public through increased awareness, communications, education and "creating stewardship opportunities for citizens and communities to play a role in our Urban Forest."

The Ribbon of Green (ROG) was an initiative that pre-dated the UFMP. The program Concept Plan originated in 1990 and Master Plan dated 1992. This inspired a 4-Phase "What We Heard" outreach, beginning in May 2017, with Phase 1: Share Your Ideas (to formulate policy). An extensive array of communication strategies, e.g. newsletters, social media, media, posters, road signs, website, were initiated to connect with and involve the 3 groups of interest – Stakeholders, the Public (via online and in-person) and the Indigenous community. These strategies continued to provide the foundation for the following phases: Phase 2: Shape Our Ribbon in December 2017 (design), Phase 3: Plan Our Ribbon in July 2018 (Plan refinement) and Phase 4: Confirm Our Ribbon in January 2019 (final Plan). The entire What We Heard report phases can be viewed at https://www.edmonton.ca/city_government/initiatives_innovation/ribbon-of-green-public-engagement.aspx

In response to the UFMP, a number of programs were developed. Follow-up questions and additional information were kindly provided by Root for Trees Greening Program Supervisor, Lisa Fialka, where indicated:

LF – Root for Trees was created as a stewardship greening opportunity for citizens. While the program itself has gone through a few iterations since its introduction, the following principles have stayed consistent in both Root for Trees programming and development.

Vision /Principles

Our vision is to have a city of trees planted by Edmontonians.

The following principles are outlined in the UFMP, and permeate our program:

1. Promote a healthy and sustainable Urban Forest
2. Engage the community in protecting and managing the Urban Forest
3. Think globally and regionally; plan and act locally
4. Use best practices, innovation, science, information and technology

Goals

We aim to establish a diverse and sustainable Urban Forest by expanding the current tree planting and educational programs. Our annual target of planting 45,000 trees aims to achieve the overall goal of doubling the size of Edmonton's Urban Forest, from 10% coverage (outside of the river valley) to 20%, by 2022.

Listed are the desired outcomes of our programming:

1. Enhanced natural and public areas
2. Attractive and well-utilized public green spaces
3. Greater public understanding of (i) the benefits of trees and green spaces, (ii) the challenges facing Edmonton's Urban Forest, and (iii) the best practices used by the City to maintain the Urban Forest.
4. Greater public appreciation of Edmonton's Urban Forest, and the benefits of naturalization in urban areas
5. Increased ownership of Edmonton's natural areas due to participation in improving specific natural spaces

Implementation

Our volunteers are from a wide-variety of ages, ethnicities and educational backgrounds and have the opportunity to sign up for either a Guided, Have Shovel Will Travel (Self-Guided) or a Special Event.

NZ – Are the naturalized areas within the City of Edmonton municipal boundaries?

LF – Yes – all naturalized areas are within the City of Edmonton.

Root for Trees planting sites are selected from areas that have been identified for naturalization. Naturalization is when the landscape is allowed to become more natural by restricting mowing, and by planting trees and shrubs that are found naturally in Alberta.

Many sites that are selected for naturalization are along arterial and collector roadways. They may also be in areas that are deemed unsafe for crews to mow (*e.g.*, an area with a steep grade). Root for Trees typically does not plant in these areas; instead, these areas may be planted by operational crews (ex. professional tree planters).

Other areas selected for naturalization are within parks and neighborhoods. These may be areas that are under-utilized by the public, or have been identified as areas where naturalization would be beneficial for the local ecosystem. Here are two examples: (i) naturalizing an area may help connect two other natural areas, thus improving the wildlife corridor, or (ii) increasing a natural area around an ecologically sensitive area to provide more protection. Many new subdivisions or housing developments may also have identified areas that will be set aside for naturalization. These are the areas where Root for Trees may plant in the future.

Before planting can begin, however, the general public is made aware of the plans to naturalize the area; other departments within the City are also consulted. Feedback is collected, and the designated area location is circulated to make sure other departments or groups don't already have plans for the area (*e.g.*, a developer may have plans to turn an empty field into a soccer pitch).

Root for Trees plants only on City property or public land. We do not plant on private land (including School Yards), and do not plant on areas managed by other levels of government (ex. Transportation utility corridors, like the Anthony Henday).

NZ – You have indicated that Root for Trees volunteers cover a broad age and ethnicity range. Can you give me a rough age range as well as examples of ethnicities which have participated to date?

***LF* –** As participant ages is not a statistic that we track, I can only provide a general age range: 4 years old – 65 years old. Generally speaking, the majority of our volunteers are adults (18 years+). We interact with a variety of groups: from corporate groups, school groups (Grades 1–12), a variety of community groups (*i.e.*, Girl Guides, Scouts), day cares, religious groups and families. Given the diverse groups that volunteer with us, ages range from toddlers to seniors. We welcome participation from anyone who is willing and able to plant.

As for ethnicities, I would say that our volunteer base is representative of the larger Edmonton area.

NZ – How do you recruit volunteers? Does the strategy vary with age range and /or ethnicity?

***LF* –** Generally speaking we do not target a specific subset of Edmontonians, with the exception of some of our Special Events (*e.g.*, Arbor Day with special invitations to Grade 1 classes). The program has grown substantially since it's introduction in 2012, largely by word of mouth, social media (rootfortrees.ca, Facebook, Twitter, Instagram) and our participation at volunteer fairs.

NZ – Is there an educational component and how is that component disseminated – online or handouts or…? If there is an online educational component, can you provide a link to the curriculum?

***LF* –** One of the projects that I will be working on in the near future is putting together a digital newsletter. The plan for this newsletter is to include a variety of educational topics and activities that volunteers can read about or partake in, respectively.

Our programming is largely operational in that, our primary objective is planting as many trees/shrubs as possible within any given volunteer shift. We do integrate varying degrees of educational content into volunteer opportunities:

(i) At guided events, the educational component is comprised of a verbal introduction to **Root for Trees** at the beginning of an event (sample script linked here), as well as some fun facts about the species that are being planted that day. We do not have specific educational handouts for events.

Guided Events criteria: generally 1.5–2.5 h in length, these are booked timeslots led by a Root for Trees Program Leader at one of our naturalized planting sites (outside of the River Valley). https://www.edmonton.ca/city_government/environmental_stewardship/naturalization.aspx These plantings are the 2nd stage in the Naturalization process; native trees and shrubs are planted to establish a healthy and diverse ecosystem. Root for Trees provides all of the necessary materials needed for a successful planting event: plant material, shovels, and gloves. The average group size is 27 volunteers.

NZ – Who is leading the guided events? Are they part of the municipal UF Management agency or…?

***LF* –** Greening Program Assistants (employed by the City of Edmonton) facilitate planting events with Root for Trees.

(ii) Self-Guided volunteer leaders are provided with a brochure on how to properly plant seedlings.

Self-Guided (Have Shovel Will Travel): designed for small groups who want the freedom to plant on their own schedule. Volunteers are given the option of the week they would like to plant, as well as their preferred self-guided planting location. These plantings are the 3rd stage in the Naturalization process; smaller native trees, shrubs and wildflowers are planted to enhance beauty, biodiversity and to provide resources for wildlife to flourish and re-establish. We will provide ~100 small seedlings and up to 15 shovels. Root for Trees Program Leaders do not participate in the planting, but do provide the Volunteer Leader with planting and safety instructions when they pick up their planting kit. The average group size is 10 volunteers.

NZ – How are the self-guided Have Shovel Will Travel events organized? Does the volunteer leader need any skills and, if so, how is that determined? Can you share the planting and safety instructions the volunteer leader receives?

LF **–** We are currently re-assessing our Self-Guided programs to determine if they are in alignment with recent changes made to the provincial Occupational Health & Safety Act. Volunteers are considered workers under this act; as such, the employer (City of Edmonton) needs to provide them with the necessary training to perform volunteering tasks safely.

I refer to the volunteer group organizer as the Volunteer Leader – this is the individual that I, and my team, communicate with. While anyone can sign-up for a self-guided planting shift, there is a minimum requirement of 2 people (so that the volunteer is not working alone), and a maximum group of 15 people.

The Volunteer Leader would then communicate with a Greening Program Assistant to schedule a time (typically during business hours) to stop by the Root for Trees office to have a quick safety/planting talk and demonstration, as well as pick-up all planting materials. They receive pre-made kits of ~100 seedlings, shovels, gloves, a planting brochure (available online in supplementary material), a couple stakes and flag tape, and required paperwork.

Volunteers then have 1 week from the day that materials are picked up to plant their seedlings in a predetermined location (selected by one of our Landscape Technicians) and drop off their planting kit with us.

(iii) At some of our Special Events, we will have a variety of rotation of 1–2 educational activities and a guided planting. For example, at National Tree Day we have (a) an exhibitor tent were external (non-profit organizations, corporations) and internal exhibitors are asked to display educational material and a hands on activity, (b) City of Edmonton Arborists facilitate a bucket truck/climbing demonstration and finally (c) a guided planting with Root for Trees. Depending on the theme of the special event, we may tailor our speech to include information about the theme (*e.g.*, pollinators, fruit-bearing trees, *etc*.).

Special Events: feature interactive activities and educational opportunities, along with a guided planting led by our Program Leaders. All shovels, gloves and plant materials are included in these events. Special events are open to the public, and advertised well in advance through rootfortrees.ca and on social media. Each event is unique in its own way, based on a specific theme or geared toward a specific audience, *i.e.*, Pollinator Planting, Family Nature Night, etc.

NZ – Is there any after care follow up that's part of the planting initiatives? What is involved with that and how is that organized? If not, once the seedlings are planted, does the Municipality take over the establishment and aftercare responsibilities?

LF **–** Great question! At this time, we do not have a volunteer-based after care program for the newly planted seedlings. The municipality takes on the regular maintenance of the seedlings.

Once guided and special event planting are complete, the sites will be visited by water trucks over the season. Here is a typical watering schedule:

Within the first 24 h; then
Within the first 7 days;

A third watering within 7–10 days;
Every 10–14 days for the rest of the season.

The water trucks will typically visit each planting site again the following year for more watering.

Our Root for Trees planting facilitators (Greening Program Assistants) also regularly visit planting sites to perform plant health checks. This way, we monitor the health of previous planting years, and the success of the different species planted in these areas. Also, if there are any invasive weeds these are taken note of.

NZ – Is there a comparable program for the individual trees along Edmonton's streets or in parking lots, etc.? How are they being cared for? Are citizen groups involved with this critical aspect of care?

***LF* –** Currently, there isn't a comparable program to ours for individual trees and their maintenance. There is, however, a Request to Plant program (outside of Root for Trees scope). https://www.edmonton.ca/residential_neighbourhoods/gardens_lawns_trees/request-to-plant-form.aspx

New Delhi, India – Give Me Trees Trust Tree Planting

I was introduced to Akshat Tyagi by an i-Tree team colleague, Alejandro Zalaya, as a result of a project outreach.

New Delhi, India's capital since 1947 and the home of the Give Me Trees Trust, is the most significant representation of an urban agglomeration in the country and, as with most of the World's sprawling urban centers, it has its share of problems, especially those that are environmentally-based.

Akshat Tyagi holds a bachelor's degree in Mechanical Engineering from Delhi Technological University. He met Peepal Baba, the founder of Give Me Trees Trust, in February 2017 during a tree plantation drive. Inspired by the tremendous work done by Peepal Baba and his volunteers, Akshat joined the Give Me Trees team. He presently works as the Head of Digital Operations and Strategy for Give Me Trees Trust. In that capacity, he started the digital team for Give Me Trees in April 2017, and began setting up the end-to-end digital platform for Give Me Trees and Peepal Baba. Over the last 3 years, he has implemented various digital operations, e.g. web development. He is also responsible for the expansion of the organization – recruiting more employees, channelizing the existing volunteer force as well as digital marketing, collaboration with other organizations and the media, collaboration with multi-national corporations for corporate social responsibility projects, etc. He actively coordinates with volunteers on social media, email etc. and was instrumental in delivering a successful i-Tree Eco project in New Delhi. Apart from working towards preservation of the environment, he is really passionate about music.

Akshat has shared the following information about this important greening initiative in New Delhi and throughout India:

NZ – What is the Give Me Trees Trust, when was it established and what is its Mission:

***AT* –** Give Me Trees Trust is the largest tree planting and preservation environmental organization in India, having planted and preserved more than 20 million trees throughout India and having more than 11,500 volunteers. We have planted trees across all states of India. We have been working towards preserving the environment since 1977 when environmentalist Swami Prem Parivartan founded the organization – 43 years and counting. The mission of Give Me Trees is to increase the native tree population throughout India.

Our focus is on planting native trees as they are the best for the environment and have a huge life span. We also extensively focus on maintenance, manuring and watering of the saplings until the trees become self-sufficient. The strength of Give Me Trees is our immense assemblage of volunteers.

NZ – Can you provide some information about Swami Prem Parivartan and how he came to found such an important initiative?

***AT* –** Swami Prem Parivartan (popularly known as 'Peepal Baba') is the founder trustee of Give Me Trees Trust. He has been planting trees since 1977, when he was 11 years old. His work on peepal trees (*Ficus religiosa*, also known as the Bodhi tree under which Guatama Buddha meditated and attained enlightenment) is so extensive (having planted more than 10.25 million peepal trees) that people and media nicknamed him Peepal Baba. His life is a story of building a large constituency of volunteers across regions who now work for the propagation and promotion of peepal trees. His insistence on preserving and serving each tree sapling for 2–3 years from the date of planting became the mantra for his volunteers. The more that people understood his work, the more that joined him. He was inspired by his school teacher and grandmother when he was 11 years old to plant trees, and that became his lifelong mission as time passed. His love for the natural world – trees, bees, birds and butterflies, rivers, forests, farms and mountains – kept him on his path of working with trees.

Peepal Baba frequently visits various schools to create an awareness among children about trees and environment.

Peepal Baba is also a renowned TEDx speaker, sharing his experiences and inspirationthroughhisTEDtalks:https://www.youtube.com/watch?v=vJgeq5nT1RQ

https://www.youtube.com/watch?v=vJgeq5nT1RQ

NZ – How does Give Me Trees work with communities to engage with and recruit volunteers?

***AT* –** <u>Tree planting and maintenance</u> – we have given green skills training in more than 6000 schools, teaching the students about how they can plant trees, maintain planted saplings etc. We also hold regular sessions at universities and institutes all across Delhi NCR.

We have planted one of the largest man-made Urban Forests in Delhi NCR. This Urban Forest is known as Uday Upvan and is situated in the Noida district of Delhi

NCR. Before we started planting, this land was a wasteland, devoid of any greenery. Within 2 years, our team along with our volunteers transformed this 5 hectare area into a lush green forest of more than 43,000 tree saplings. Birds and animals started to come back to the area, including several species such as the house sparrow, which is nearly extinct in Delhi. This forested site has been visited by members of World Bank, United Nations and many other large organizations. The transformation of this barren land into a lush green Urban Forest within a span of just 2 years is widely regarded as model technique to create an Urban Forest.

We work with our volunteers in many ways:

Community outreach – we hold regular plantation drives across Delhi NCR, inviting more and more volunteers to come and plant trees with us. We also hold several awareness campaigns across Educational Universities and residential communities on how to plant trees, take care of the planted saplings including water management, waste management, composting.

Composting – for more than 40 years, we have used kitchen waste to create high quality compost, that is used to provide organic nourishment to the soil and thus, the tree saplings that were planted. We have trained more than 50,000 households to make compost using kitchen waste. More than 4000 households are regularly giving us their kitchen waste that we use to make high quality compost, which is then used at our tree plantation sites.

Training and Workshops – we offer regular workshops on tree plantation, tree maintenance, waste management, water conservation, making nests for birds, making eco bricks, etc. We offer these workshops at schools, colleges, universities, corporate houses, residential societies and many other places.

Give Me Trees had a major plantation drive last year at our urban forest, inviting several schools to participate in planting saplings with us. With the help of community engagement, we ended up planting more than 4200 tree saplings in a single day. Please have a look at the YouTube link of the event below:

https://www.youtube.com/watch?v=GsEVOQOf3A0

Rooftops for growing tree saplings – we have set up huge nurseries for growing tree saplings on rooftops, providing trees for planting and greening rooftops at the same time. When the saplings get big enough, we take them from the rooftops and plant them in the soil. Thousands of households across Delhi NCR help us to use their rooftops for keeping and maintaining young tree saplings.

Seed collection – we encourage people to collect seeds and to give them to us. We use these seeds to generate new trees. For example, during summer, people eat a lot of mangos in Delhi NCR. We run several campaigns during this time, asking people to collect the mango seeds and send them to us (either in person or through courier) which we then use to plant more mango trees.

We also encourage people to donate seeds to us through our social media campaigns.

https://www.youtube.com/watch?v=IiXaEkWSyUY

NZ – The educational component is very strong in your initiative and it would be very interesting to have more details on exactly what the green skills training curriculum consists of, not only for planting the saplings but their after care. Can you please share this information as that's what would help readers replicate this model initiative?

AT – The **'Green Skills' training** consists of the following components:

(a) We start with sensitizing people about the current situation of environment in India. Some topics include – the tree percentage cover in Delhi and what percentage is ideal, native trees vs. non-native trees (why native trees are better), how trees serve as magnets for water, myths surrounding trees etc. We don't only speak about the environmental benefits of trees, but the spiritual aspect as well.

(b) We also teach people about how trees are related to our mental health and how spending time with them can reduce stress levels.

(c) After the introductory awareness session, we teach the attendees the methodology to plant saplings.

(d) We also teach them how to maintain the saplings once they are planted (watering frequency, mulching, composting etc.). Our organization has a special focus on making sure that all the saplings we plant are maintained for a minimum period of 2 years, so that they become self sustaining. We have a tree survival rate of more than 90%.

(e) We also give detailed training on several techniques and methods:

(i) Using their rooftops as nurseries for saplings:

We encourage them to put young native tree saplings (species like Peepal [*Ficus religiosa*], neem [*Azadirachta indica*], banyan [*Ficus benghalensis*], mango [*Mangifera indica*] etc.) on their rooftops which have tremendous benefits, instead of ornamental plants which provide little benefits. They take care of the young saplings for a period of 6 months to 1 year, providing them with water and compost, till the time they are big enough to be planted in the ground. Once the saplings are big enough, the volunteers either plant the saplings near their homes, or give them to us so we can plant them along road sides, road dividers, in schools, universities etc.

(ii) Compost generation:

We teach them how to make very high quality compost using kitchen waste. A large proportion of the household waste in India is organic kitchen waste, which can be used to make high quality compost using simple techniques. This compost can then be used as manure for the young saplings, or even for gardening.

Apart from in-person training, we also upload instructional and educational videos on YouTube for making compost, tree planting, tree maintenance as well as many other techniques and tips. Our founder Peepal Baba explains how to make compost using kitchen waste in this video:

https://www.youtube.com/watch?v=fDBlNNMHICQ

(iii) Waste management:

We use milk packets, delivery packets etc. as saplings bags (to hold young saplings for a period of 6 months to 1 year) until they are big enough to be planted in the soil. We teach the training attendees how they can use easily available waste at home to hold saplings.

Please have a look at this video:

https://www.youtube.com/watch?v=mZW-sH7sl9s

(iv) Mulching:

We teach the attendees how they can use household waste like coconut shells for mulching, which is very important for the maintenance of the planted saplings. Mulching helps prevent summertime loss of moisture from the soil into which the tree saplings have been planted. This takes on even greater importance, given the very high temperatures Delhi NCR experiences over the summer as well as the lack of water availability. Please have a look at this video in which our colleague explains about mulching:

https://www.youtube.com/watch?v=j4HKv7Ikogw&t=5s

(v) Making bird houses:

We also hold several workshops on how to make bird houses. Please have a look at the below video:

https://www.youtube.com/watch?v=p7bFnrJJX8s

(vi) Eco bricks:

We teach people how to use plastic waste to create eco-bricks. We use these eco bricks to create structures at our tree plantation sites. Please have a look at some the photos of the structures made out of eco bricks at our plantation sites.

https://drive.google.com/open?id=1-2HDM-QOYgSQ2INrKzmu_H2sVTblrsI-

(vii) Creating new trees using root extraction:

This training focuses on a very important technique – creating your own trees instead of buying them from nurseries. We extract roots and use them to create completely new saplings.

https://www.youtube.com/watch?v=jj3q9wepBA8

(viii) Planting trees using stem cutting:

We also teach people how to create new trees for planting using stem cutting in the below video:

https://www.youtube.com/watch?v=_VEJd4OiyMk

(ix) Bio enzymes:

Bio enzymes are enzymes made from organic waste that are very good for not just the growth of saplings, but also for purifying water bodies. We hold workshops and trainings on how to create bio enzymes.

Our Work with Schools

The aim is to provide students with an opportunity to learn in an environment that will help them connect with nature, make an impact at their level and make them more aware of and responsible for the environment.

Our Eco-schools programmes have 4 components, namely:

Biodiversity:
1. Open green spaces
2. Tree plantations

Waste:
1. Waste management
2. Composting
3. Waste Segregation

Water:
1. Water Management
2. Drinking Water
3. Rainwater Harvesting

Healthy Living:
1. Creating vegetable gardens for students and teaching students urban farming.

NZ – What is the approximate amount of time volunteers have to allot to training?
AT – The training length actually depends on what the attendees are interested in learning about and how much time do they have available. We tweak the training accordingly.

For some people, it can take 3–4 months if they attend sessions every weekend. For others, it can be shorter as well if they intern with us full time. We have had people come over from many countries across the world (Egypt, Brazil, France, Germany etc.) who have stayed with us full time for 30 days. They take this knowledge back to their home countries and apply it there.

We also have 1–2 day events where we train people on the very basics of tree planting. These events are not very technical, but rather give a general overview of tree planting.

Of course, the attendees don't perfect the skills within a few days or months. But it gets them started on the path.

NZ – You mention plantation drives – how do you do this? Do you have partners such as the media and, if so, what information do they disseminate?
AT – We organize at least 2–3 plantation drives every week. We hold plantation drives in the following ways:
1. through social media (FB, Instagram);
2. coordination with residential societies and planting trees with residents in their localities;

3. through WhatsApp groups;

4. through our corporate social responsibility (CSR) partners – we hold massive plantation drives every month with the employees of our CSR partners;

5. coordination with schools – we have worked with more than 6000 schools since 1977. We hold plantation drives either at the school premises, or the students come with their teachers to get trained and plant trees at various other plantation sites;

6. through coordination with companies – we work with employees of several companies interested in holding trainings either at their office premises, or at other sites. To date, we have worked with more than 54 partner organizations;

7. through media coverage – several media houses, like NDTV, Brut India, Aaj Tak, DD News, Hindustan Times, Lok Sabha TV have covered our work. They help to spread awareness about our work;

8. through word of mouth.

NZ – When you go into a new community, how do you make connections to potential volunteers? Do you communicate with community leaders and, if so, how do you find them?

***AT* –** When we go to plant trees in a new community, we sensitize the local people about the importance of trees in their immediate surroundings and how the trees would benefit them. As soon as people understand the importance of trees, they willingly help us out. Once we have planted the tree saplings, we involve the local community in taking care of them for the next 2–3 years, till the time the trees become self sufficient. The local community then maintains several of the trees planted in their neighborhoods and motivates several others to plant trees.

We have more than 40 active plantation sites within Delhi NCR. At one of these sites, within 2 years, we transformed a wasteland (which was a dumping ground for garbage and human waste) near a residential society into an Urban Forest of more than 2500 trees. The residents of the colony played a major role in maintenance of the planted tree saplings – the preservation of the planted saplings, protecting them from unwanted human intervention, grazing cattle, watering them regularly and keeping the saplings healthy.

https://www.youtube.com/watch?v=krv_VCOHepY

We find the local leaders with the help of the community members. Also, once we start with our tree planting work there, the community leaders automatically get to know about us.

This video shares the experiences of a resident of the society who now takes full care of this Urban Forest.

https://www.youtube.com/watch?v=IHtjsSy1lO8

The media house NDTV has covered our work in the below videos. Please have a look at these as well.

https://www.youtube.com/watch?v=om8CL6OwaCc
https://www.youtube.com/watch?v=iH9k20DYrto&t=11s

York Region, Canada – Adopt-A-Stream Tree Planting Program

The York Region of Canada is located in the south eastern part of the Country in the southern part of the Province of Ontario and the Greater Toronto Area. The region's geographic footprint has an extensive percentage of naturally-occurring hydrologic features, many of which are within the boundaries of towns and villages as well as cities.

I was introduced to this program and Michelle Bourdeau by Adrina Bardekjian of Tree Canada. Michelle Bourdeau is the Program Manager, Green Infrastructure, in the Natural Heritage and Forestry Division of the Regional Municipality of York (Ontario, Canada). She holds a BSc in Forest Conservation and is an ISA Certified Arborist. The Region's Green Infrastructure program is responsible for implementing the Region's Greening Strategy, the York Region (strategic) Forest Management Plan, and other strategic Natural Heritage and Forestry initiatives. The Greening Strategy operates within four pillars: Enhancement & Rehabilitation, Outreach and Education, Environmental Land Protection and Preservation, and Leadership, Innovation and Knowledge. The Region's award-winning Greening programs are delivered in collaboration with other organizations and departments, resulting in over 100,000 trees and shrubs planted annually, over 90,000 residents reached each year, and over 1376 hectares of land protected (through land purchase) and enhanced since 2001. These activities contribute to the Forest Management Plan goal of increasing woodland cover to 25% and total canopy cover to 35% by 2031.

Michelle has kindly shared aspects of the York Region's Adopt-A-Stream Tree Planting program, as follows:

NZ – Can you state the program's mission as well as its goals? When was its inception and how was it decided to involve residents?

***MB* –** Direction for the program comes from York Region's Greening Strategy. In 2016, the Region formalized the Adopt-A-Stream Tree Planting program, which had been in existence since 1995, as well as other stream, wetland and riparian rehabilitation projects. Through the York Region *Adopt-a-Stream-Tree Planting* program, school and community groups have the opportunity to help restore forest cover and habitat in-and-around streams near Regional roads. This dovetails well with our delivery partner's (Ontario Streams) mission: the conservation and rehabilitation of streams and wetlands through education and community involvement.

Goals include:

- to promote the Program to residents through various opportunities and media;
- to foster a coalition of stewardship and education interests to participate in the Program;
- to foster a sharing of resources, technical capabilities, knowledge, experience, and training in order to increase naturalization initiatives on private and public lands; and
- to raise awareness of watershed protection, stream naturalization and restoration, as well as local aquatic species-at-risk.

NZ – Are the streams you select for adoption in immediate proximity to a built environment – a City or a village?

MB – Volunteers and groups select adoption site(s) close to their neighborhood, club meeting spot, business or school; these locations are easily accessible without travel and their proximity to the community increases the likelihood of participants already possessing some knowledge about the site. With the landowners consent, groups can adopt rivers, streams, riparian areas or wetlands on crown land, municipal land, conservation authority property, or private property.

NZ – How are the streams you adopt selected? Do organizations or citizens contact you or…?

MB – York Region works with a delivery partner, Ontario Streams (a local registered not-for-profit, since 2019, with technical expertise), to identify suitable sites on Regional or municipal properties. The important factor is ensuring that the site has a water body that can be improved. Streams can be in close proximity to both urban and rural areas.

In 2018, the program re-directed focus to streams on Regional and Municipal properties, moving the focus away from Regional roadsides due to safety concerns and challenges with roadside tree planting.

NZ – How do you reach out to prospective participants? Does the partnering organization do the outreach to citizens? If so, how is that done – through what mechanisms? If not, how are participants recruited?

MB – We use media advisories and releases, social media campaigns and advertisements. Our delivery partner also agrees to promote the Region's role in the Program through recognition in promoting the Program and at outreach events, and through social media, media releases, website and print materials.

NZ – Do you have a trained cadre that you reach out to asking them to participate in upcoming activities and providing dates, location, etc. or…? Do you use email or is there a Facebook page or…?

MB – Programs and outreach are led by Ontario Streams staff and volunteers. Ontario Streams staff promote the events via emails in a mailing list, social media and other public events sites.

Some groups are return participants and direct contact is maintained with them (i.e. schools, scout groups and select businesses). Combined Insurance Company of America is an example of a business that has a repeat commitment. They will participate in their 10th annual tree planting event in Spring 2020.

NZ – How would you describe participant demographics in terms of age range, ethnic diversity, gender, etc. Do you ever get relevant professionals volunteering?

MB – Age range, ethnicity, diversity and gender are not tracked.

NZ – Can you describe any training or educational component that is offered, e.g. why riparian zones are so important, how the plants within these zones manage the potential environmental extremes… Is there material available online? Is there a script or curriculum you would be willing to share?

MB – Ontario Streams offers an in-class component for school groups participating in the program, related to stream health, riparian zones and tree planting. They also offer an excellent resource online entitled "Ontario Streams Rehabilitation Manual" https://www.ontariostreams.on.ca/Resources-Publications

York Region's Adopt-A-Stream Tree Planting Program is under the supervision of Krystal Athanassiou. Krystal, with a BS in Environmental Studies, has worked for Ontario Streams since 2015, beginning as a Habitat Technician then promoted to Project Biologist, and finally to her current position as Field Supervisor. Krystal's responsibilities include project management, fund raising, reporting, data collection and interpretation, and stream and wetland restoration, in addition to the facilitation of several of Ontario Stream's outreach and education programs. In the field, Krystal supervises the implementation of Ontario Streams' field projects and provides guidance and training to junior staff members.

"Through York Region's Adopt-A-Stream Tree Planting program, participants have the opportunity to learn about the Region's role in greening our communities, the methods and benefits of tree planting and how to restore healthy riparian corridors. In 2019, Ontario Streams implemented four Adopt-A-Stream planting/stewardship events where 73 residents planted 409 trees and shrubs."

NZ – How do you keep past participants engaged? Is there any follow-up during the establishment period as well as for aftercare? Do participants remain involved with the adopted area and, if so, how?

MB – In 2020, there is interest from new and returning participants. Ontario Streams has the responsibility for monitoring their restoration projects. There are no currently offered follow-up tasks for the adopted area, unless the same site has more space for projects in future years.

NZ – Is there any involvement from local schools? Have you done this work with students and, if so, what age range and was this experiential education which addressed specific curriculum components?

MB – Yes, there is involvement from local schools as school groups are one of the key audiences of the program and their participation is experiential. Students are typically 10 years + when participating to ensure that they can actively participate in a safe manner.

NZ – Are there any related activities, presentations or the like so that these stewardship endeavors are shared with the larger community to inspire them to get involved?

***MB* –** Ontario Streams offers a suite of stream, wetland, and riparian rehabilitation initiatives, including species-at-risk habitat rehabilitation, and community-based environmental stewardship programs throughout the Region which support the Region's Greening Strategy.

In 2019, Ontario Streams planted 3529 trees and shrubs on 25 riparian sites and engaged with 295 residents; and they attended 3 outreach events reaching 250 participants.

Some restoration sites are in Redside Dace habitat, a federally recognized species at risk. The importance of pollinator species, especially Monarch Butterflies, is highlighted in every Adopt-a-Stream Tree Planting event. In preparation for the 2019 planting season, Ontario Streams collected local pollinator-friendly seeds and planted them at various locations throughout York Region.

Engaging senior citizens is recognized as a supporting initiative, including interest from senior citizen community centers and homes but none have expressed interest in participating at this time. Ontario Streams continues to promote to York Region seniors to take part in their programs to encourage healthy living.

Ontario streams provides outreach staff for some regional events to promote the work that they do with the Region and educate attendees at events about the importance of healthy streams and riparian areas.

York Region resources:

https://www.york.ca/wps/portal/yorkhome/recreation/yr/outdooreducation%20/!ut/p/z1/jZBNb8IwDIb_ynboEWwyKBG3qGNLC6hI20TJZWqhXxrUVUg3bb-e0GmHSVCw5IOtx_b7GhREoKr4s8xjU1IV72y9Vu67L559KWcYhEPuocBQBGzMcfqKsGoBvBACQd0y3wGo7vXBtQPWAdMLb5GDqmNT9MoqI4ioMVsinW6bTWv1zgpR7SrGhq4ceBigDDn6T-Pl6JHLAbrsCjD7A7rV5jtKfh8rquSBW1k6zVKd6n6jbbswpj5MHHTQSvs4ZU79De0dPDdS0MFA9J-Eev8W_cyzF7-nku8vcX8EbfTTlQ!!/dz/d5/L2dBISEvZ0FBIS9nQSEh/#.XlEq8C2ZMvB

Zagreb Croatia – The Green and Blue Sesvete Tree Planting Project

I first learned of Zagreb's tree-planting/beautification project through my monthly proGIreg newsletter. proGIreg, an acronym for productive green infrastructure for post-industrial urban regeneration, also known as Nature for Renewal, is a project funded by the European Commission under its Horizon 2020 program which began in 2018 and is expected to conclude in 2023. Its focus is Nature-based solutions and a philosophy of "Co-creation for urban regeneration. Together, local citizens, governments, businesses, NGOs and universities design the nature-based solutions and make them happen" https://progireg.eu/the-project/

In the project's initial "four front-runner cities" – Dortmund DE, Turin IT, Zagreb CR and Ningbo CH – "eight different nature-based solutions will create productive green infrastructure that not only helps improve living conditions and reduce vulnerability to climate change, but also provides measurable economic benefits to citizens and entrepreneurs in post-industrial urban districts." https://progireg.eu/nature-based-solutions/background/

ProGIreg's project in Zagreb has partnered with a project that began in 2015 entitled ZIPS – the Green and Blue Sesvete project. ZIPS focus is "a healthy living environment for its inhabitants and availability of quality public benefits based on environmentally friendly, sustainable and organized growth". ZIPS one employee who began as a volunteer – Marijo Spajic – is a Landscape Architect by profession and now coordinates the proGIreg project. In addition to Marijo, ZIPS has over 50 members that come from all segments of civil society, and from the entire city district of Sesvete, all professions and different ages (dominated by young people) and all of their activities are exclusively based on volunteer work and contribution. In support of the project, ZIPS has been joined by the City Office of Strategic Planning, providing communications with the public via its public forum, ZgForum.

Marijo's involvement with ZIPS and the project is an inspiration in affecting change and initiating connections. "I assumed an NGO (ZIPS) could be a great driving force for change in my hometown. You still need a great support of people in politics, but if you do not complain about the bad situation AND show by an example how it could be done, not much is going to change. Politicians usually don't listen to requests for these kinds of changes until you shout really loudly, especially in the rocky Balkans. Also, I love working with people. People do not realize they can make a change, so our biggest job is to educate them how it could be done. We are a young and tentative democracy, still evolving to understand that we as small people can make a change. People usually expect someone else will do it and murmur (or just give up and leave the country). You don't change the landscape by waiting for politicians, you do it with people and for people".

Marijo has kindly agreed to share project information with us:

NZ – What was your vision for the project? What did you hope it would accomplish?
MS – The role of the NGO ZIPS in this project was to connect local actors in the Municipality of Sesvete in order to raise awareness of the natural, cultural and all other resources of the Sesvete area. The target groups were all citizens of Sesvete who we wanted to sensitize to the necessity of environmental protection and the need for sustainable development. The activities of the NGO are directed primarily towards education of the population. Particular emphasis was placed on the youngest population – children and young people. The indirect beneficiaries of this project are all residents of the City of Zagreb, but especially city district Sesvete, to make a place where they live and recreate and make it more comfortable and healthy.

NZ – How did you decide on the location for the project?
MS – The reason we chose this particular street is because it is located in an industrial area. In this area, we can find a large number of marginalized groups and a large

number of high school students who walk daily from Novi Jelkovec to the center of Sesvete. By planting trees, we wanted to create a better image of the area and new microclimatic conditions for a large number of users who are walking along this street. It is important to note that there was not a single tree in the profile of the street before this tree planting action.

NZ – Besides ZIPS, did any other Zagreb organization participate in the project as a partner?

***MS* –** We contacted Hrvatske šume (Croatian Forestry) to ask them for a donation and partnership in the planting which they provided, including the donation of plants from their nursery. Then we found another partner to promote us and be a part the project – Zabavni radio (Fun radio), which is a local radio station.

NZ – Given that most of the planting was on publicly owned land, what was required for you to do that officially?

***MS* –** At the end of summer (there is no one in Zagreb during July and August), we started the bitter love relationship with bureaucracy. First, we needed an approval from the City district of Sesvete, after which we got the support from the City Office of Strategic Planning (partners in proGIreg project) and Zrinjevac (city-based firm for planning and planting). Afterwards, we had to get a permit to plant, based on a plan. Since I am a Landscape Architect, I did the design in coordination with the Masterplan and the planting plan, but also had to get all the blueprints from all the firms connected to the infrastructure (water, electricity, gas, telephone, Internet etc.). Unfortunately, there is no common server to get the data, so I had to go from one firm to another, as if we were building a 20 floor-high building. Still, we managed to hit an electric pole and one optical cable since it wasn't in the blueprints. Surprisingly, no one was hurt in the process. At the same time, I was designing the pamphlets and posters and we were promoting it all over the place, inviting people, sports clubs, other NGOs etc.

We still didn't get the permit, but we decided to dig anyway, knowing it would arrive at some point. The whole October was quite rainy, so we barely managed to get the surveyor out to locate the planting spots with GPS (all donated services from Mijo Lažeta, Geomjer). Then we got an excavator from the local firm željko Andrijanić (also donated), and it was dug out quite smoothly. We also decided to contact all the local firms in the area where we were planting. Sesvete has a Master Plan, but parts of the infrastructure corridors are owned privately because the district still hasn't purchased everything. So we asked them for permission to plant on their property where the trees should be according to the Plan. They all agreed and even offered to help in planting. One milk factory – La Log – even gave us food and drinks and NEL voće provided fruits. One private landscaping firm, Foton promet, helped us with planting the trees and they also purchased the trees from the nursery and another, Zeleni svijet – gave us soil and cleared the land and Agroproteinka provided work support. Sesvete District provided the water truck and staking materials

Fig. 5.8 Zagreb Green and Blue Sesvete Tree Planting Partner Poster. Marijo Spajic

as well as post-planting establishment irrigation. We put all the firms on our posters (Fig. 5.8).

NZ – Besides partnering with Zabavni radio, how else did you promote the event?
***MS* –** We made t-shirts to promote the event (partially a donation from Monaco) and cotton bags. Ostvarenje, a local NGO that does art with kids took the bags and the markers we bought them and they were drawing all the protected species of plants

and mushrooms on the bags (on the day of planting and a few days before). We gave the t-shirts and the bags to the participants, so they keep promoting us around.

The local community was informed through modern methods of communication with citizens, such as social networks and websites. We informed the general public through media such as: Zabavni radio (local radio station), Radio Marija (national radio station), Radio Sljeme (local radio), EKO Zona HRT (national television), Z1 (local television), 24Sata (national newspapers), Sesvete-danas (local web portal) and many others.

NZ – How did the planting participants know exactly what to do? Was there training or information provided on site or online before the day or…? If available online, would you be willing to share any of the materials you created for the project?

MS **–** The participants had "experts" next to them, who showed them how to plant the trees (landscape architects, horticulturalist, people of Zrinjevac and Foton promet). We divided the planting area into a few areas, so these experts walked around and helped people to plant the trees the right way. The expert supporters were organized by Zrinjevac. There weren't any online courses because we figured there wouldn't be much use of it because we would have the experts participating.

NZ – Please describe the planting day(s). Who participated and how many trees were planted?

MS **–** Because of everything that was planned – a movie premiere and presentations – we decided that this event was more than a tree planting, it was a "festival". The "festival" took place from 23rd to 26th October and, in addition to the tree planting included the premiere at a local college, Narodno sveuČilište, of the movie "Man and the Tree", by Dalibor Platenik. The producer usually shows this movie only on festivals, but he decided to make an exception since he loved the idea so much. All donated. We also planned presentations but had to move them to November. On October 26th, 2019, in cooperation with the Sesvete City District, the City of Zagreb, and with the support of Zrinjevac (City landscaping firm), we carried out planting of 162 trees of different species in a range of sizes from 1 to 4 m as well as 40 shrubs, up to 0.4 m.

A large number of participants, around 250–300, starting as young as 1 year old, joined our action, with a great emphasis on the children's population, sports clubs, hikers, high school students, Landscape Architecture students, other associations and many other interested citizens. The realization of this project will raise the awareness of the local population, especially among the young population, about the potential of the open and green areas in the city district of Sesvete, as well as the

Fig. 5.9 Zagreb Green & Blue Sesvete tree planting. Sesvete danas, Borko Samec; Marijo Spajic

importance of protecting and regulating the environment in order to serve all citizens of both the district and all residents of the City of Zagreb (Fig. 5.9).

NZ – How was the project received by the local Sesvete community and the City of Zagreb? What kind of feedback did you receive?

MS **–** After the project, we were all over the place. People of Croatia are usually quite cynical, but this was one of the most positively accepted projects I have witnessed. Everything about it was perfect – the vibe, the people, the sunny day... We did have a problem with one NGO that operates on a soccer field owned by the city (they didn't allow us to plant there, even though it's not up to them – they didn't want to collect the leaves from the field), but that was just marginal. We got about 300 new likes on our Facebook page in a couple of weeks and people viewed us far more positively.

NZ – Is there any planned follow-up with the project's planting participants – something to keep the momentum going? Who is providing the establishment protocols or the aftercare? If citizens, how is that being organized?

MS **–** There is going to be a follow-up project similar to this one. In April or May of 2020, we plan to plant an Arborertum of St. Francis about 1 km from this place. The land, about 40.000 m^2, is owned by a Monestary, which is willing to open it to public use (it already is). Aftercare is done by the City landscaping firm Zrinjevac, since they're the only ones allowed to do it.

NZ – Do the schools involved have school yard gardens? Has ZIPS thought about how to keep the youth that participated engaged in greening and caring for their environment? Is there anything in school curriculum which can have this sort of experiential education as a component?

MS – Yes, all of the schools in Sesvete have a yard garden, so we provided them with small trees. Kindergartens as well. And yes, of course we thought about it, that's why we invited all the kindergartens and schools to join us. A month ago I was teaching kids from a local kindergarten what a Green city would be. I think there is school curriculum that is experiential. There already was something when I was a kid and I think it has improved in the meantime. Usually it depends on the good will of the teachers.

NZ – How would you summarize what you accomplished with this project?

MS – We managed to bring the proGlreg project and 1,5 million Euros into Sesvete, but this was far more positively accepted than that. People like to see the change and be the part of it. The proGlreg project is far more exhausting in terms of papers, meetings etc., but that's not something people see. What they saw and felt in just a few hours was the positive change. We rebranded the street most people didn't think about a few months ago, even though it's only 100 m from the District center.

http://www.sesvete-danas.hr/vijesti/veliki-broj-sesvecana-u-subotu-sadio-stabla-oko-sljemena-galerija-fotografija-i-video-10074?fbclid=IwAR2NFEs4w7lQk7y5TlyMlVeKs-smYj41amkxHHDYhXP2qX3xpxBMJA5EQiM

Post Planting Care and Long Term Maintenance

The forests are dying, the rivers are dying,
and we are called to act.
To return Earth to harmony is to restore
the harmonious principles within ourselves
and to act as responsible caretakers
– to save the forests and the waters for future generations
Dhyani Ywahoo

Greenpoint Brooklyn New York USA – Greenpoint Tree Corps

The origins of the Greenpoint Tree Corps date back to 2016. That's when Acacia Thompson, a resident of this north Brooklyn neighborhood, started a group called North Greenpoint Neighbors. The group's original mission was twofold – advocating for future park space for the north Greenpoint community and building a network of tree stewards to care for street trees. After a year, this effort evolved and was concentrated on launching a grassroots street tree care collective with an initial outreach to the neighborhood's licensed Citizen Pruners. Thus, in February of 2018, Greenpoint Tree Corps was born.

Acacia was joined by Sarah Balistreri as the two driving forces behind the Greenpoint Tree Corps initiative. Further momentum was provided by Joe Ferrari, a neighbor and plant shop owner and from the Greening Greenpoint project, funded by the Greenpoint Environmental Community Fund and managed by NYC Dept. of Parks & Recreation. Additional support was afforded by Partnership for Parks' Fellowship Academy, a 6 month program that trains park groups and street tree stewards in organizing, grant writing, programming and varied other grassroots organizing needs.

I learned of this group and its bottom-up street tree care initiative as a result of following the Greenpoint community's greening endeavors since my involvement with the community's battle with Asian Longhorned Beetle (ALB). It is an important lesson in possible outcomes from tree disasters the likes of ALB and the Tree Corps project is a credit to Acacia, Sarah and the entire Greenpoint neighborhood.

Sarah has kindly provided a great deal of detail, as follows:

NZ – How do you recruit your street tree stewards?

SB **–** 1. Email: we have a good email list of a few dozen Citizen Pruners who live in Greenpoint. We acquired this list with help from Trees New York.

2. Social media: we post about events on Facebook and Instagram, which sometimes drums up volunteers. There's a very active North Brooklyn Crowdsource Facebook page, which helps us reach a local audience of engaged neighbors.

3. Word of mouth: Acacia is great at reaching out through her apartment building and parent networks to get people involved.

Fig. 5.10 Greenpoint GO GREEN Festival Poster. Greening Greenpoint

4. Partner organizations: many of our events have been in partnership with other community organizations (Greening Greenpoint, North Brooklyn Parks Alliance, Trees New York), and they've helped us connect to volunteers (Fig. 5.10).
5. Flyers: old fashioned flyers around the neighborhood do work!

Fig. 5.11 Greenpoint Tree Corps tree pit maintenance. Greenpoint Tree Corps

NZ – What's the age range of your street tree stewards?
***SB* –** Depends on the event. We've had people of all ages help steward trees. Kids and mulch always become fast friends, so we try to get them to help with laying mulch down in tree beds. One of our most regular volunteers is a retired nurse who can clear a tree bed of weeds faster than anyone I know. (Fig. 5.11)

NZ – How often do your stewards go out to work on tree beds?
***SB* –** Our aim is to hold at least one event per month in the summer/fall.
NZ – Who supplies the mulch? Can you request composted wood chips as opposed to shredded mulch which compacts too easily?
***SB* –** Partner organizations have been incredibly supportive with supplying items like mulch, tools, and technical assistance. North Brooklyn Parks Alliance (formerly Open Space Alliance) has provided mulch, as has Greening Greenpoint.

NZ – Is the Tree Bed Care 101 brochure the only training that stewards receive for caring for tree pits or do you use other strategies?
***SB* –** Depending on the event, we have different kinds of stewards that require different kinds of training.

For example, when we hold pruning events, we only recruit NYC-licensed citizen pruners, since they are the only ones allowed to prune NYC's publicly-owned street trees. They have all completed the Citizen Pruner course and have basic pruning and tree care knowledge. I have been lucky enough to attend numerous tree pruning outings with Arborists, and therefore, I've developed knowledge beyond the citizen pruner course that I try to use when guiding volunteers on Greenpoint Tree Corps pruning days.

Other events have focused on tree bed cleanup, mulching, planting perennials, or even just tree identification to open our neighbors' eyes to the street trees around us. In these cases, we give instructions as needed to ensure volunteers don't overuse mulch or damage tree roots when cultivating the soil.

NZ – How do you disseminate the brochure? Is there an online mechanism? Do you have it as a handout through neighborhood retailers or…?

***SB* –** I created the Tree Bed Care 101 guide for an event a few summers ago. I chose a tree on a busy shopping street with the aim of transforming the tree bed over the course of a day, all while stopping passers-by and engaging them in conversations about tree bed care. Sadly, the chosen tree bed was full of rock and clay and it turned out to be impossible to plant in it. Nevertheless, we cultivated the soil and added compost and mulch so that—one year later--it is in much better shape. We also engaged with dozens of community members that day, so it was a success in many ways!

After that, we handed out the Tree Bed Care booklet at various events where we had a table, e.g. Kingsland Wildflower Festival https://www.greenroofs.com/2018/09/19/3rd-annual-kingsland-wildflowers-festival-in-greenpoint-this-saturday-september-22nd/ and at our own volunteer events. We share the brochure with anyone who stops by our table.

NZ – Do you do any outreach to neighborhood residents/businesses who are repaving their sidewalk to make the tree opening larger or to install permeable paving, rather than cement?

***SB* –** This was one of the major efforts of the Greening Greenpoint project. I don't have exact stats, but they enlarged many tree beds throughout the neighborhood. We do have plans to create a tree care guide specifically for businesses.

NZ – Have you requested that tree planting contracts for Greenpoint require the installation of 20 gallon TreeGators to facilitate watering and the necessary slow release of the water?

***SB* –** No, we haven't. It seems that the Parks Department really struggles to hold their tree planting contractors accountable, but I don't really know the ins and outs of that topic.

NZ – Have you thought of requiring or requesting any of the following:

- that new sidewalk installations allow for the clustering of trees being planted curbside or the installation of root paths or soil trenches to allow curbside trees to connect underground? Cluster plantings in particular provide trees a greater accessible volume of open soil while taking much less space overall. Connecting tree's roots through root paths will keep trees healthier, provide a greater accessible soil volume and reduce infrastructure conflicts.

- that the soil substrate under new sidewalks be a structural soil, such as CU® soil, if the other side of the sidewalk is an open landscape such as a front garden? Structural soil is an excellent conduit for moving tree roots to a larger accessible soil volume which they will find in the open landscape while reducing root/infrastructure conflicts.

SB – Love these ideas! Our mission is focused on stewarding trees in Greenpoint and educating our community members about tree care, so we haven't channeled our energies toward influencing citywide Parks Department policy in this way. This is definitely something we could consider in the future, though.

Preservation and Protection – Advocacy

Never doubt
that a small group of thoughtful, committed citizens
can change the world.
Indeed, it's the only thing that ever has.
Margaret Mead

Given that a great deal of participatory activity is actionable, that requires an investment of energy to provide momentum and propel the actions. It is community advocacy that energizes involvement and this is often initiated by a leader(s) who has a commitment to the relevant goal, an understanding of how to inspire interest and an ability to bring interested stakeholders to an open table.

Portland Oregon USA – Share-It-Square Community Advocacy

Share-it-Square is located at Sherret Street and 9th Avenue in the Sellwood Neighborhood of Portland, Oregon.

Share-It Square was the first community-initiated and community-built project in Portland. A group of Sellwood neighbors began building an unauthorized gathering space in 1996 for a cost of $65. The reaction of a Portland city official "That's public space. Nobody can use it." did not deter the community. The neighbors chose to move ahead with transforming the intersection because a crossroads can be a gathering place where people can come together. "In America, our great archetype is the main street, which is not really a center. It's just a flow. It's a movement

Fig. 5.12 Sherret Street Share-It-Square composite. Ashley Bonn, Yvonne Schumacher Strejcek

corridor, and you have to yell across the street because there isn't a place in the middle. There isn't a social commons that you can attain and occupy."

Since 1996, changes and improvements have been ongoing. The intersection has been a place-making endeavor which celebrates community and the urban ecosystem in the form of the wonderful trees that frame the square (Fig. 5.12).

Local residents were primarily responsible for Share-It-Square's design. During the first project, the neighbors who lived around the intersection came out on the weekends to design and build all the structures. Each of the four corners of the Square contains a community-oriented amenity, which includes a cob dispenser for the monthly neighborhood newsletter The Bee, a 24/7 T(ea)-Station, and the "Little Free Library", an innovation which has now gone viral, the Share-It-Square Community bulletin board, a children's playhouse/craft corner and cob benches. Structures incorporated into the design are made of natural and/or reclaimed materials, such as the cob benches, and produced by the local residents.

After the initial installation was complete, organizers surveyed their neighbors and found that an overwhelming majority (over 85% in each case) felt that crime had decreased, traffic had slowed and communication between neighbors had improved.

Since then, new projects abound. People have built saunas, put in gardens and helped each other paint their houses. While Americans typically move every 4–7 years, not in Sellwood. The families that live right around that intersection remain because people want to live there, clustering around it, having kids or bringing their kids, increasing the number of children in the neighborhood. There is more shared childcare and more adults interacting with kids on the street. The intersection was repainted for an estimated, resident-generated $500. One resident designed and built an earthen oven, shaped like her Australian Tree Frog which is used for pizza parties. A number of residents organized a collection to raise money to purchase fruit trees, huckleberry bushes along with other edible plants so that when planted, they would turn the neighborhood into a "fruitopia" for everyone's indulgence.

While just an illegal tea house originally, the project has reframed the city government's acceptance of the need for public community space and endeavors such as Share-It-Square, passing an Ordinance enabling other communities to implement similar initiatives.

Share-It-Square's community initiative has not only inspired other similar projects throughout Portland, it also instigated spinoffs of the original effort within Sellwood. This has resulted in under used and/or abandoned public spaces being turned into community gardens, forest farms and ecological restorations.

Victoria Park Western Australia – Urban Forest Strategy – Grassroots Advocacy for Strategic Planning – One Citizen's Journey

Heather Johnstone, with a background in textile design and education, thought of herself primarily as a housewife, raising a daughter and contending with life and family issues. In an effort to make her life well-balanced, she took to walking daily, spending an increasing amount of time in the nature that was near her home as well as a residential park.

"During these daily walks, I began to notice a pattern of tree removal for residential infill development. As I also live adjacent to an industrial zone with a cement batching plant (60 m from my home), I became increasingly interested in the environmental impact of tree removal on my and the community's health and well being.

In 2015, my daughter asked me why all the parks were so far away from our house (nearest is 1 km) and I began talking to the Water Corporation (Government agency responsible for water infrastructure management) and the Town of Victoria Park with the idea of transforming a drainage basin into a park. I hand-drew some plans and talked to my neighbors and the cement plant and everything looked very promising. When it turned out that neither the Town, the Water Corporation or the cement plant had the processes in place to fund and/or manage the project, it all fell in a heap and I felt pretty sad for failing my daughter. My lack of professional qualifications and experience were also a factor. By this stage however, I had already begun to read about water sensitive urban design, urban ecology and landscape design."

The 2016 Lathlain Park development project, which cut down 98 mature trees with 8000 m^2 of canopy valued at close to $1,000,000 to make way for expanded football grounds, provided the necessary motivation to take ownership of canopy loss.

"I was able to use my established relationship with Town staff and, in collaboration with Peter Melrosa and Luana Lisandro of Vic Park Trees (formerly Victoria Park Urban Tree Network of which Peter was the founder), take on the role of community liaison between residents and Town staff."

To be better prepared for the role she was adopting, Heather went to Melbourne to talk with their Urban Forester, Kelly Hertzog, returning with a bunch of resources from Melbourne City Council. "I was also attending as many free workshops as I could and trying to understand what skills I would need to be useful. Training has included asset-based community development, co-design, marketing, GIS mapping, urban ecology, planning, arboriculture and volunteer management." It also turned out that involvement with a community group was therapeutic and beneficial for her mental health.

The evolution from liaising with Town staff to writing Victoria Park's Urban Forest Strategy https://www.victoriapark.wa.gov.au/About-Council/Council-documents/Urban-Forest-Strategy was facilitated through conversations with the

Fig. 5.13 Victoria Park Volunteer Contributions. Heather Johnstone

prior head of Parks and Victoria Park's CEO, both of whom were interested in leveraging support for asset-based community development. With their support, Heather was able to recruit enough people to enable her to develop and draft the Victoria Park Urban Forest Strategy (UFS) together with community members Jim Mellor and Vicki Caulfield (Fig. 5.13).

In addition, the community's thoughts were engaged through the Your Thoughts website page where citizens could offer their ideas on all the varied Urban Forest Management Strategies being explored:

https://yourthoughts.victoriapark.wa.gov.au/urban-forest-community-forum
https://yourthoughts.victoriapark.wa.gov.au/urban-forest-community-forum/forum_topics/discussion-space

The UFS was endorsed in September 2018 by the Victoria Park Town Council https://www.victoriapark.wa.gov.au/Around-town/Environment/Urban-Forest-Strategy and a budget implemented in June of 2019. The Town Council's website has hosted varied aspects of the UFS endeavor including frequently asked questions https://yourthoughts.victoriapark.wa.gov.au/urbanforeststrategy/faqs

The Implementation Action Plan (IAP) co-design workshop took place in March of 2019. The resulting Victoria Park Implementation Action Plan https://www.victoriapark.wa.gov.au/About-Council/Council-documents/Urban-Forest-Strategy-Implementation-Action-Plan was endorsed in September of 2019. Attendees at the IAP workshop received a stipend of $500 Australian for the day which was donated to set up a related microgrant project https://www.vicparkcollective.com/single-post/2019/06/09/Urban-Forest-Strategy-Microgrant as well as in support of refugees.

The reader can follow this bottom up initiative through its partnerships:

- Vic Park Trees, a community-based advocacy group in Perth, Western Australia. Through their website, the reader can access a whole host of grassroots projects https://www.vicparktrees.com/home and related Urban Forest Strategies from other Australian cities https://www.vicparktrees.com/urbanforeststrategy
- Vic Park Collective, a community-based organization https://www.vicparkcollective.com/about where one can discover their Streets Ahead approach to collecting community input as well as action initiatives https://www.vicparkcollective.com/streets-ahead as well as their partner organizations – Millennium Kids https://www.millenniumkids.com.au and the Australian Urban Design Resource Centre https://www.audrc.org
- The reader can also access the Collective's involvement with Victoria Park's UFS https://www.vicparkcollective.com/urban-forest-strategy

An important part of grassroots participation in planning is visioning both for the participants as well as for the intended decision makers . To that end, the Victoria Park citizens initiative easily synchronized with research work being undertaken at the Australian Urban Design Resource Centre which was then realized through practical applications developed by the Collaborative Place Design (Fig. 5.14).

Fig. 5.14 Victoria Park UF Strategy Visioning workshop. Dr. Anthony Duckworth

Dr. Anthony Duckworth has kindly shared information about the Victoria Park Urban Strategy collaboration.

NZ – How did you become involved ?
***AD* –** We had been producing interactive models to help with community engagement, specifically for new housing infill policy. One of the significant findings, which also ended up in planning policy, was the desire to retain and enhance trees. Not specifically for increasing canopy cover, although that was qualified, but participants also identified trees with a variety of benefits such as neighborhood character, sense of place/memory, birdsong and insect homes, links with micro-climate, amelioration of building bulk and greenhouse gas sequestration. The community group Vic Park Collective became aware of the way trees had formed an integral part of this process and policy. So when they were asked to undertake the Urban Forest Strategy, they approached us to consider how our work could be adapted to community engagement for such a project.

NZ – How did you see your role in this particular grassroots initiative?
***AD* –** Our role was to try and mediate between the objectives of the community engagement process, time and resources available and the design, production, application and facilitation of an interactive model. Our role was to provide the means for learning, invite participation, provide an inclusive engagement environment and enable input and invention from the community (who had great knowledge regarding trees, soils, etc.).

NZ – What information did you require to facilitate accomplishing the tasks you have described?
***AD* –** Principally, what the strategy's objectives were and how the process could align with that. We needed to know the specific urban contexts that were being investigated (e.g. suburban, industrial, town center) and then various targets for canopy coverage etc. Also we needed to know the numbers of participants and the types of engagement forums that we would be attending.

The reader can access the Collaborative Place Design's website to view other community-engaged endeavors https://collaborativeplacedesign.com as well as a public talk on the projects and their process https://youtu.be/lMC3kuPEj9E

This extensive collaboration and the resulting combined initiatives that have been undertaken by this Urban Forest grassroots community are facilitated through the group's Facebook page Vic Park Urban Forest Supporters, a quick and easy way to reach the entire group. While the existing Urban Forest Strategy and the Implementation Action Plan will not solve all of Victoria Park's Urban Forest issues, it's an excellent example of bottom up advocacy – its evolution and the results to date are clearly a work-in-progress as the group will keep the pressure on the Town Council to expand the initiative and make it more all-encompassing. It is clearly a win for the Town, its urban trees, the landscapes they populate and thus, all the Town's residents – flora, funga, fauna and human.

Beyond Trees – Reconnecting with the Land

The soil is the great connector of lives,
the source and destination of all.
It is the healer and restorer and resurrector,
by which disease passes into health, age into youth, death into life.
Without proper care for it we can have no community,
because without proper care for it we can have no life.
– Wendell Berry,
The Unsettling of America: Culture and Agriculture

Community Gardening

Community gardening on public land dates back to the nineteenth century, addressing many issues beyond just the need to grow food in all the varied places and spaces in which these gardens appeared around the globe.

- community gardens/forest farms are like trees – existing in community, nurturing community members, sharing resources, improving environmental and human health and supporting a dynamic sustainable ecosystem;
- they are like the forest in that they rely on the cooperation of everyone within the association;
- they cover all the bases of participation – advisory, management, experiential education, actionable endeavors and advocacy;
- they provide opportunities for diversity in the form of cultural variables. Since we all harken back to a geographically oriented agrarian existence, there were cultural distinctions in how that agrarian existence played out. Those cultural distinctions offer experiential education opportunities and help break down barriers between people of different backgrounds;

- most importantly, they keep the land green in a positive, beneficial and inclusive manner – quintessential Urban Forest.

Mattariya Community, Cairo, Egypt – Sustainable Community Gardens Owned by Successful Communities

Mattariya is a Cairo community in which 80% of the City's informal settlements are built on previous agriculture land and more encroachments are still happening on the existing agriculture lands on the periphery of Cairo. This transformation of agriculture land led to a very dense urban pattern of houses stacked together from three sides and very narrow streets, with almost no access to open or green spaces.

Environmentally, these informal areas have the least amount of green space per capita (generally Cairo is averaging at 1.4 m^2/capita... in some informal settlements it reaches 0.3 m^2/capita).. and thus, informal settlements are more vulnerable to the impact of climate change (UHI, heat waves, sand storms).

Socially, there is a lack of gathering venues. Friends and families usually gather at the entrances of their building, which are usually confined spaces, disrupted by passing cars and motorcycles, making it an unpleasant space in which to socialize.

Economically, informal settlements are home to low – middle income strata who are usually looking for means to improve their economic status. They are impacted by the continuously increasing food prices, causing them to depend more and more on calorie-dense food and gradually abandoning healthy choices of fruits and vegetables.

While at the 2018 World Forum on Urban Forestry, I had the good fortune to attend a presentation given by Abdallah Tawfic, who is from Cairo and an Architect by training. He is presently working as a Technical Professional at the Participatory Infrastructure Project (PIP), the German International Corporation (GIZ) and previously as a Researcher/Project Manager at the Research Institute for a Sustainable Environment (RISE) at the American University in Cairo (AUC). Abdallah was presenting on behalf of a community development project, "Sustainable Community Gardens Owned by Successful Communities", funded by GIZ and developed, and implemented by RISE. I knew immediately that I had connected with a model program, not just for its specific location but conceptually.

Abdallah was kind enough to respond in depth to my pointed questions, and his reply follows:

NZ – What is the focus of the project?

AT – The focus of the Sustainable Community Gardens Owned by Successful Communities project was mainly environmental. The AUC tried to tie that together with food production, improving aesthetics of densely populated urban areas, and creating hubs for community social interaction through the creation of 5 sustainable urban community gardens in the informal area of Mattariya in Cairo. The 5 selected locations are semipublic spaces (two schools, two public entities (Library and Administrative entity) and one NGO) (Fig. 5.15).

Fig. 5.15 Cairo Egypt Mattariya community gardens. Abdallah Tawfic

> **NZ –** How did the idea for the project come about and how was it developed and funded – did you have any partners from local government, from within the community or professionally; how did you recruit participants from the community; was there a criteria for who could participate and how many people are participating?

***AT* –** The project was a grant from GIZ. RISE – AUC had been invited to submit a project proposal in a call focusing on the Environment and Climate Change mitigation and adaptation for Cairo's Informal settlements. The project was led by RISE-AUC as the lead Applicant, and we were accompanied in this project by a sub-applicant, Alashanek ya Balady – AyB (translates as: "For you my Country") – a local NGO that works on the training and capacity building of informal area community members. We have also partnered with the administrations of the targeted locations through Memorandums of Understanding (MoU) where we focus on role identification (we, as implementing arm, GIZ as donor, community members as the grant beneficiaries). We have also partnered with local authorities on a district level, specifically the Urban Upgrading Unit (UUU), as well as the Educational Directorate under the district of Mattariya, who supported us, especially with access to public entities, licenses and approvals to work with those public entities and schools. Their members have also requested to attend the training of the community gardens teams, which was welcomed by the project team, adding to building different capacities of the members of the local authorities.

The project went through 5 phases on a course of 2 years. Those phases were:

1. Selection of 5 locations: It was not easy to select who would be a part of this relatively small project compared to the overall scale of this informal settlement. The GIZ had previously done a Participatory Needs Assessment (PNA), which was a guiding tool showing where we could focus our efforts. We also did several visits to different potential locations, and we had our "Checklist" tool , which is a technical and social questionnaire that helps us properly evaluate the location and the people working/living there.

After surveying different locations we did an internal evaluation workshop and came up with our decision for the previously mentioned locations. Our checklist focuses on the technical conditions of the surveyed space, including water accessibility, drainage and, most critically, roof condition including insulation, load bearing capacity, parapet height (safety), accessibility of the roof, wind speed and direction, solar exposure/obstruction. The social checklist is based on conversations with the local community, asking them basic agriculture questions to see if there were any previous urban agriculture attempts and evaluate their interest to be part of the project.

2. Create Community Garden Teams: As previously stated, after selection, we have signed an MoU with each of the selected locations' administrations, and we requested from each entity the creation of a community garden team ,consisting of 5–6 members working/living within their entity, and at least 50% of them should be women. Those members will be responsible for the daily activity needed for the community garden, and who will be trained by RISE and AyB.
3. Design the Community Garden: this step was done through a coordinated effort between RISE staff (who are mainly architects, urban planners and landscape designers), and the 5 teams. We (RISE) created simple 3D models of the designated spaces and there was a back and forth coordination with the community, to make sure that the design fit to their needs, and to include them from the early decision making process to reinforce and build on the value of ownership.
4. Train the Community Garden Teams: After settling on the design, RISE started two parallel actions, tendering and procuring the materials needed, as well as training the community members on Garden Implementation and Management over a course of 5 days, as well as Social and Green Entrepreneurship over a course of another 5 days. The aim of the second training was to provide the community members with the basics of marketing, business model canvas and the means on how to sustain the activity in their garden, how to sell their produce and/or expand their gardens if they have the possibility and the support.
5. Implementation of the Gardens: The next step was to implement the gardens. This happened in a joint effort between RISE staff and the targeted community members. A collaborative implementation process was adopted – all 5 teams joined their efforts to implement the 5 gardens – to increase the hands-on and technical experiences of the community members through repetitive implementation, as well as increasing community bonds between the 5 teams and creating a small network of urban farmers.
6. Post implementation: Upon handing over the gardens to the community members, RISE staff made periodical visits to check on the management, operation and maintenance of the gardens. Those visits have decreased through time, as the project was reaching its end and we wanted the community members to gradually handle their gardens on their own. We have created a "Best Practices Manual", a 75 page booklet, documenting the project phases, and including technical data that should guide the community members after the project ends. We have also created open channels of communication with the community members (ongoing) through

WhatsApp groups, where the members can freely share their successes, failures, stories, and inquiries with the rest of the Community Garden team members.

NZ – Has interest in the project grown within the community and in other communities – has it served as a model?

***AT* –** During implementation, we were approached by neighbors asking questions about what we are doing on the roofs, and if there might be a possibility to extend our project to their roofs. We had a very interesting experience with a neighbor from an adjacent building who spotted our daily work during implementation. We had a water supply loss during one of our implementation days, and this neighbor offered to give us a water supply by throwing the plastic hose to her and connecting it to her kitchen faucet. Her reply to our thanks was that she wants to support us because we are providing her with a good view from her window.

We received a request to share copies of the Best Practices Manual with the educational directorate, who are planning to use this booklet during the agriculture lessons in all of the Mattariya schools. We did not map the impact beyond the boundaries of the district of Mattariya, but we were approached by developers who want to adopt the concept and implement it in other informal settlements.

NZ – What has the project enabled within the community?

***AT* –** I think the social impact for this project was remarkable. In one of the schools, the teachers have created a rooftop farming team composed of students who were also responsible for garden activities under the supervision of the teachers. Other teachers have been using the roof for outdoor activities for their students. The trained team members are now training the youth and passing on knowledge and experiences and creating young stewards.

Financially, the project might be facing challenges in terms of sustainability. The trained members produce an amount of crops that are not enough for further improvements and operations of the garden, given also the limited space of the garden. Financial support is usually required from the school and educational administration, who have more urgent topics that need financial support (basic school infrastructure).

NZ – How does the project reflect cultural traditions and norms?

***AT* –** During the garden design process, community members requested space for their weekly gatherings. Some members are still using the community garden to gather with colleagues and friends for morning tea. We also know that school teachers are hosting regular dish parties serving traditional food, were they compete on their cooking skills and share photos and stories of their events with everyone. In addition, some teachers use the space for meditation and spiritual practices. They said that they enjoy reading Quraan sitting by the planters under the shade of the pergola.

St. Gallen, Switzerland – Community Gardening – The UrbanesGrünAtelier (UGA) Labor und Atelier für eine grünere Stadt (Laboratory/Studio for a Greener City)

I learned of this creative and inspiring grassroots project through a casual LinkedIn conversation with Florim Sabani, a Swiss colleague who is now located in St. Gallen in the eastern part of Switzerland. The project began in July of 2019 and is situated behind the St. Gallen freight station

NZ – How did the idea of creating a *community garden* first get discussed?
FS – Well, in the beginning there was just myself and two friends discussing the pros of having green spaces in the city. It's not only the municipality but also the residents who are responsible for their greenspaces. People tend to hand over responsibility to the municipality or the government and are consternated if nothing happens. The other part of this discussion in St. Gallen can be condensed to the following question: To whom do the public spaces belong and how do we create and shape them?

So we decided to put action into this whole discussion by creating and shaping our own small project which could then grow from there.

NZ – Does St. Gallen have many Schrebergarten or is this something that is not happening in the city?
FS – Since St. Gallen is called *die Stadt im grünen Ring* (the City in the Green Ring), there are quite a lot of *Schrebergarten* (little gardens with little huts; usually rented) also called *Familiengärten* in relation to the size of the city. However, the UrbanesGrünAtelier is not to be confused with a Familiengarten or Schrebergarten. In a Schrebergarten, the community as a community is usually not very strong. Every participant does his or her own thing and they do not really create a community feeling. In the UGA this is one of the highest goals.

NZ – How many people were organizers and what are their backgrounds?
FS – Today, the board or society's committee (*Vorstand*), consists of 7 people: a Forester/Urban Forester, a social worker (President), a chemist, a gardener, an industrial designer, a programmer and a communication consultant. But we don't see this as a work which needs to be professionalized. We accompany the process that the participants create – we do not steer them.

NZ – How did you decide on *where* and how did you approach the SBB (Swiss Federal Railways) for its use?
FS – It is very tricky to find "abandoned" spaces in the city. So we had to take what we could get. And it was a piece of luck. There was this overgrown greenspace that everybody knew of but nobody seemed to care about and we found out who was responsible for it. The person we spoke with was willing to hand over the lease to us so we contacted the person responsible at SBB and told him about our plans and

ideas. He said that as long as the SBB doesn't have anything to do with it (work hours-wise), it sounded like a good plan to him.

NZ – Is the SBB donating its use to your organization?
FS – The SBB is not donating at all. We pay the normal price for it (250 Francs per Year and per 2/3 Hectare). They just paid for a part of the reinstatement costs, since there were very old, almost rotten wooden barracks that needed to be torn down.

NZ – Did the land have any contamination issues?
FS – Not that we are aware of although it is possible since its near a railway track. We soon should take some soil samples actually. But the vegetation and the soil horizons do not indicate that there are bigger issues.

NZ – Once you knew you could use the space, how did you get the word out? How many people responded immediately?
FS – We formed a *Verein* (Association) and informed the municipality, the existing *Quartierverein* (Neighborhood Association) and the immediate neighbors. And since we have contact to media, we contacted them as well. Those were the first steps. Everybody – the municipality included – responded within a very short time and was and is very positive.

NZ – Is there any criteria for how much space someone gets and what they can grow?
FS – We always look from case to case, from project to project. It actually doesn't matter what they grow (as long its not some sort of invasive species) but it matters how they grow. Its forbidden to use chemicals and synthetic fertilizer. They get the space they need, as long as other projects are not affected in a negative way. We have 3 categories of prices – it gets more expensive the more space the specific project needs.

NZ – Is the space open 24 h and are there any restrictions/requirements as to how the space can be used?
FS – Yes, the space is open 24 h. So far, this works out perfectly fine and people respect the neighbors and so on. We don't want to restrict things that do not need restriction. The spaces can be used for whatever they want – it just has to be following our very vaguely formulated statutes. Once we get a *Projektantrag* (Project proposal), the board decides what is possible and what is not.

NZ – One of the articles indicates that projects must be *nature compatible*. What does that mean? What do you require be presented in order to know that the project will conform? Who has the knowledge to make that decision? What are some of the specific criteria?
FS – Nature compatible: the project can not under any circumstances affect the environment (socially and naturally) in a negative way. If it is compatible, everything is possible. We – the board – have the necessary knowledge on many levels to make that decision. The project group needs to present a description of their ideas and then we discuss that directly on site.

– We don't have specific, formulated criteria. It's just common sense actually. I know, this is very "guerilla", but it applies very well in the situation we have at the moment. We all agree that overregulation is not necessary at the moment. Following common sense is enough for now. Maybe we have to change that in the future, as this grows. I think about these criteria quite often and I would be happy to have your thoughts on that.

We want to be very low-threshold. because we want to include people with no or almost no "green thumb". It is funny that a lot of people say they have no idea of gardening or greenspaces but if they follow their instincts and get some background information, a lot is possible.

NZ – Can you describe some of the projects that have been approved and who the participants are?

FS – *Urban Apiary*: two brothers from a nearby district are both working in a non nature-sector but they are very fascinated by bees. So they decided to start a training at the *Bienenzüchzerverein* (Association for Beekeepers). This training lasts 2 years and is primarily practical instruction.

- *Hens*: a group of eight people (all neighbors of the UGA) incubated eggs and started to keep 12 chickens this year
- *Mushroom growing:* A project to grow *Pleurotus* mushrooms has been submitted. The project is approved but he hasn't started yet.
- *Gardening with refugees:* The HEKS (Hilfswerk der Evangelischen Kirchen der Schweiz/Aid organization of the Protestant Churches of Switzerland) started a project called *neue Gärten Ostschweiz* (New Gardens of Eastern Switzerland): https://www.heks.ch/was-wir-tun/heks-neue-gaerten-ostschweiz We work with refugees who bring a lot of experience in gardening (since they are/were farmers in Eritrea). They do not need guidance on how to grow their crops. This is an ideal "integration" by working together.
- *Meeting place*: People wanted to have a meeting place in the garden. So we helped them to plan and implement that work. Now we have a place with 2 fireplaces, a stonewall to sit on and so on.
- *Company 200 test site:* An environmental engineer experimenting with Salix for protective structure (see also https://www.company200.com/English/)
- *Wildflowers*: A flower shop owner wants to grow flowers in the city (yeoon.ch)

NZ – How are artists and their works being incorporated?

FS – We don't have concrete art projects so far. It is in the planning. One artist wants to shape figures out of tree trunks. but this project hasn't started yet. We are in contact with an artists' collective and hopefully this or next year, we can start with one project.

The only art project that is already finished is a graffiti art project. A graffiti artist asked us if he could draw on our tool-storage room. We asked to see sketches, we then said yes, and voilà!!! I see the UGA also as projection surface for wishes and ideas with a laboratory character.

NZ – Do participating artists also have to show that their work is *nature compatible*?

FS – Their work within the UGA has to relate to nature topics. Whatever they need for their installation has to follow the rule, that it does not negatively affect the site or the social environment. Art is a broad topic. We also think of theater in the green and so on. But so far, these are only ideas.

NZ – Do you plan to offer any training to interested citizens?

FS – We wanted to start with 3 workshops this year but we had to postpone them due to the corona situation. There would be the *BeeNews-Seminar* where people can get in touch with the work around a beehive. Also a *Wildkräuterseminar* (Wild Herbs Seminar) where we give information, collect herbs and plants and cook with them on our fireplace. The third seminar would be the *Rasenmähen mit der Sense Seminar* (Mowing by hand seminar) where people could learn how to mow with a scythe. And of course, we give people assistance with their own project not necessarily connected to the UGA (writing concepts, budgets, etc.).

One thing I want to start next year is a *Baumspaziergang* (Tree Walk) where I lead people through the city and tell them about trees in urban areas, the threats and opportunities.

NZ – Have you thought about involving students and, if so, how would you do that?

FS – Yes I have thought of that. But since St. Gallen doesn't have a University for Natural Science, we would have to spread out to Winterthur or Zürich. But we stay in contact with the University of Applied Social Sciences in St. Gallen. I really hope something moves on from there.

Engaging with Students

The environment we live in, be it urban or rural, can feed curiosity and offer the opportunity for exploration into realms filled with magic and mystery – a fascination feast. Our participation in the sharing of all of that with students will help form the adult to be. Engaging students in the world around them can afford them access to that magic and mystery and help them embrace it as the stewards we need them to be. Imagine knowing that by planting a tree, you can start an entire Universe, what a child's delight to have created the possibility of that myriad of life – seen and unseen, known and unknown. And, through participation, we can help students learn about and steward this remarkable universe that they have initiated so that they are not just starting a Universe, they are GROWING it.

And, in starting that Universe – that myriad of life – and growing it, one can explore the importance of diversity. Biodiversity – ecosystem, genetic, species and intraspecific. Diversity which equates with variety, dynamism, difference. Diversity, be it amongst flora, funga, fauna or people, is an essential ingredient for a flourishing ecosystem and for life in balance.

Oakville Canada – Be an Oakville Tree Hero Tree Planting Program

The Oakvillegreen Conservation Association has, since February 2019, facilitated a tree education and planting program for Oakville students and youth entitled BE AN OAKVILLE TREE HERO.

The program, through an array of classroom Powerpoint presentations combined with experiential, hands-on activities, introduces participants to their native tree species and the importance growing these species has for the environment as well as for Oakville's residents, who depend on the extraordinary benefits these trees provide.

The following programmatic information has been provided by the Oakvillegreen Conservation Association (OCA):

BE AN OAKVILLE TREE HERO "engages students in environmental stewardship through learning about the benefits, challenges and threats our Urban Forest is facing and through participating in a native tree seed growing activity. The action-oriented approach of the program includes a number of outdoor tree stewardship events and participants are invited to plant locally-grown native tree seedlings from local seed sources back into the community and therefore, increase Urban Forest canopy and diversity" (Fig. 5.16).

Fig. 5.16 Be an Oakville Tree Hero composite. OCA

NZ – How did the idea for the program come about?
***OCA* –** We believe in the power of Environmental Education for building greener and stronger communities!

Engaging students in environmental stewardship has always been a priority for us. Having worked with Oakville schools for many years, we saw the need for a program that would support curriculum but also provide an opportunity for students to experience urban ecology right at their school and community, and to realize they can make a difference! We were hoping to design a program that would ultimately help youth gain appreciation of and protection for our valuable Urban Forest.

BE AN OAKVILLE TREE HERO included it all – Informative, Inspiring, Hands-on, Interactive, Curriculum Connections, Local Knowledge and Action, Community Partnerships.

NZ – How many schools have participated and how have they responded?
***OCA* –** To measure outcome for the period February 1–June 30, 2019, we performed an evaluation survey (Table 5.1) with teachers participating in the program. We also looked at metrics (Table 5.2), social media engagement, case examples, quotes, photos and other informal feedback. Many of the related social media posts had higher engagement than average. Through the program, we were very successful in engaging various school groups and multiple grade levels as well as community youth groups, such as Scouts groups, a youth leadership group and a youth hockey team. For this period, we have been able to reach 1230 students!

Table 5.1 Evaluation survey results offered to participating schools in the Be an Oakville Tree Hero Program

Satisfied with the program delivery	Program being age appropriate	Relevant to Ontario Curriculum	Increased participants' environmental awareness	Students were more encouraged to protect the urban forest
100%	100%	80%	100%	100%

Table 5.2 Students, presentations, tree walks, trees started from seed, tree planting events, and trees planted. Be an Oakville Tree Hero Program

# of students reached	# of school presentations delivered at schools	# tree walks led with students and youth groups	# of trees started from seed at schools	# of tree planting and stewardship events held with school groups	#of trees planted
1230	31	21	383	5	200

Since the beginning of the 2019/2020 year, we have had an additional 300+ students participating in the program.

BE AN OAKVILLE TREE HERO has struck a very positive response within the Oakville schools. Both teachers and students particularly like the fact that the focus of the program is on the local Urban Forest, the threats it faces and the actions that could be taken locally. There is a big demand and interest in the hands-on planting/ seed ID activity and the outdoor tree walk.

NZ – What activities do you do with the students?

OCA – BE AN OAKVILLE TREE HERO includes the following modules:

- Call All Tree Heroes – Students learn more about Oakville's Urban Forest and unique tree species via in-classroom PowerPoint presentations combined with interactive learning and were inspired to take action, enjoy, care for and protect trees around their school and community.
- Grow your own super tree – Engages students in environmental eco-action by participating in a Native TREE SEED GROWING activity and interactive presentation about the life cycle and types of trees. Seed growing materials and supplies are provided. Students learn how to care for tree seedlings, monitor growth and understand the benefits of growing a native tree from a local seed source. For this activity, we have been supplying native tree seeds for the following species: red oak (*Quercus rubra*), bur oak (*Quercues macrocarpa*), sugar maple (*Acer saccharum*) and eastern redbud (*Cercis canadensis*). Sugar maple seeds have been collected and prepped for 2020 Spring plantings. Monitoring and comparing growth has been a key learning feature for the students as well as our staff. The "slow" rate of growth, while in student care, emphasizes that larger trees cannot be taken for granted ~Lesson: trees take generations to grow into a mature tree canopy!
- Join the Tree Walk Challenge – Students participate in an Oakville-wide TREE WALK CHALLENGE that expands their tree knowledge. They explore the Urban Forest around their schools and neighborhoods, led by an Oakvillegreen staff person, learning tree identification skills, interesting facts about native trees, threats that invasive species pose and how to become local community tree stewards.
- Plant to Enhance Oakville's Urban Forest – Students participate in a tree planting event in the community as well as invasive species removal, mulching and tree care for schoolyard trees. Students continuing to care for their young tree seedlings at home are invited to Oakvillegreen's public community tree planting event.

NZ – Why do you think this is important and what does your agency hope to accomplish?

OCA – The project benefits Oakville's Urban Forest because students are involved locally in environmental stewardship by planting and caring for trees, and they gain a better understanding of the benefits of trees and their care and maintenance. Trees planted adjacent to schools or in nearby parks contribute to the enhancement of the urban ecosystem and provide additional environmental benefits. This program works well in connecting indoor presentations to outdoor eco-actions, and providing valuable interconnected learning opportunities and outdoor time for students. The programming is modified according to students' grade level and interests and allows for student-driven action components, supported by Oakvillegreen, such as invasive species removal, litter pick-up or mulching. It is a wonderful opportunity for schools to establish or strengthen their partnerships with community organizations such as Oakvillegreen.

NZ – What are the next steps and where do you see the program going in the future?

OCA – For future projects, we would seek potential partnerships among high schools with greenhouses, so seedlings could benefit from longer growth times before planting out. For additional program continuity, a monitoring protocol for already planted trees could be developed and distributed to participating students so they can continue maintenance and citizen-science monitoring as part of their environmental stewardship activities. There is a potential to create instructions and incentives for students/schools to implement a school wide Mulchfest – could add value to students' learning and promote ongoing tree care maintenance principles.

With the Fall season, we have been modifying the program and including seed ID and collection, stratification and seedling transplanting. Additional resources will be developed to support work associated with protecting Ontario's natural tree heritage by collecting seeds from native tree species.

With the completion of our initially funded program period, we have been looking for continued funding… ensuring the continuation of this valuable program and its long-term benefits.

NZ – Is there anything else you would like to add?

OCA – To support our programming at schools, an array of educational learning resources have been developed that are also useful for future programming: presentations, planting guides as well as species information and aftercare instruction handouts.

Anelia Tichkova, a representative from the program, was kind enough to provide the following additional information in response to specific questions:

NZ – What is the age /grade level range of the students who are in the Tree Hero program?

AT – Pre-school (age 3–4) to High School (age 16–17)

NZ – Since we all have an agricultural heritage somewhere in our ancestry, greening activities can often help to break down barriers and erode perceived differences. What are the demographics of the schools that are participating? Are any of the participating students from the indigenous community? Does the participating student community include any recent immigrants?

AT – Oakville has a diverse student population with dynamic changes in demographics in the past decade, since our part of the GTA is growing faster than any other Region in Canada. I do not have exact data on the demographics of the participants but there were definitely multicultural groups (e.g. Faith Communities, Halton Multicultural Centre) including recent immigrants and students with diverse ethnic and cultural backgrounds.

NZ – How have Oakville's adults responded to their children's tree-oriented initiatives? Have you found that the stewardship lessons are rubbing off on the parents? Do parents/adults ever attend the Urban Forest planting events?

***AT* –** Parents, teachers and other adults involved have been very supportive and many of them attended our events, such as tree walks and tree plantings. Due to the program and its social media presence, we have also been able to increase our networks.

NZ – Has the official managing entity embraced the program and its results? Does anyone from the municipality's Management agency ever join with the students for their seeding/planting days?

***AT* –** We work closely with the Town of Oakville and have been collaborating on many initiatives and programs. We normally discuss and together choose the sites for all planting and stewardship projects, and align our plans in advance to ensure safety procedures and to coordinate activities with the forestry department. On a planting day, we would have the Town deliver mulch and when needed, properly dispose of any invasive species. We have received in-kind donations, such as soil, pots, tree material.

NZ – Would you be willing to share any of the referenced resources such as the planting guides, aftercare instructions, etc? If you have these online, a link to them would be wonderful.

***AT* –** We are happy to share our handouts but would appreciate if you would credit Oakvillegreen when they are either used or referenced. https://oakvillegreen.org/resources/

Padua Italy – The Woodland Close to School Project

I met Thomas Campagnaro as a result of our mutual attendance at European Forum on Urban Forestry annual events. Thomas is a post-doc research fellow at the Department of Land, Environment, Agriculture and Forestry of the Università degli Studi di Padova. He has a PhD in forest ecology and management. His current research includes management and conservation of natural and semi-natural habitats, impacts of invasive alien tree species, Urban Forestry, and landscape ecology. Thomas' colleague, Simone Iacopino, is a post-doc research fellow at the Department of Land, Environment, Agriculture and Forestry of the Università degli Studi di Padova. In his PhD, he studied the ecology of invaded woodlands by the American tree, *Robinia pseudoacacia* and the drivers of its spread. His ongoing research is focused on plant biodiversity, the impacts of invasive alien species and Urban Forestry.

Thomas and Simone have graciously shared information regarding a school project in an Urban Forest successional woodland in Padua:

NZ – Can you provide a bit of background for Padua and its embrace of Urban Forestry. Many cities in Europe do not have an Urban Forestry management concept but Padua does and I would like to put this into context with this school project.

***TC/SI* –** Padua is located in North East Italy in the Veneto Region. It covers an area of 93.03 km^2 with around 211,230 inhabitants. The municipality has a record of all municipal green spaces and of all trees that are under their management. The municipality, through its own funding as well as co-funding by the European Union for projects and collaborations, has been trying to increase citizens' attention and involvement in the Urban Forest (e.g., "Urban Green Belts", a part of the Central Europe Interreg project – a European Union cohesion policy program. https://www.interreg-central.eu/Content.Node/UGB.html

There are a wide variety of green space types such as small parks, playgrounds and roadside plantings but there are also many, relatively small abandoned lots where small forests have developed after spontaneous plant succession (Sitzia et al. 2016).

NZ – Who provided support for the project's creation and implementation?

***TC /SI* –** The opportunity to work on this project was offered by the *Informambiente* office of Padua municipality dealing with environmental education in our city. Our project "The woodland close to school" was presented and offered to the primary and secondary schools of the municipality together with other education projects on environmental issues. The number of classes taking part depends on the demand of schools' teachers and financial budget set by the municipality.

NZ – Can you describe the environment a bit – you are calling this woodland adjacent to the school *Spontaneous Woods* – was this originally a green open space that has gone to succession or was it developed and then cleared and then went into succession or...?

***TC /SI* –** Based on the information contained in the above-referenced study entitled "Novel woodland patches in a small historical Mediterranean city: Padova, Northern Italy", the area in question was a small abandoned lot which went into succession, resulting in a small forest. The fact is that, in many cases, these forests rich in non-native plant species may have less importance than the Ecosystem Services benefits such a spontaneous woodland affords the surrounding community.

NZ – What is the age range of the students involved?

***TC /SI* –** 6–11 years old

NZ – Is this project affiliated with any curriculum requirements for a particular area of study, such as Environment or was it an extra program that was developed?

***TC /SI* –** This is an extra program during the normal school hours for the students. All teachers (eight in total) involved in this program have a Masters degree related to environmental science and most of them are carrying out their PhD. The activities were planned and developed by an environmental association (Associazione Istituto Ecoambientale) for whom the teachers carried out the activities.

NZ – What type of information was covered in the lectures and workshops to prepare the students for their hands-on experience in the woods; was the information offered touching on such things as what makes a tree a tree and how its ability to grow is impacted by the urban environment? Also, did it discuss how important trees and their associates are and why?

***TC /SI* –** An introductory lesson was held with the aim of preparing the students for the forest visit. Power point animated presentations and videos were used to present "What is a Forest". This was done by enabling students to suggest plants and animals that can live in a forest. An additional activity was to highlight why the forest is *our friend* by pointing out the wide array of Ecosystem Services benefits it can provide.

After the visit to the forest, students produced a simple herbarium by using leaves collected by the operators. Students were invited to comment on leaves' smell, form and texture. Finally, species were identified through an adaptation of the "Guess who" game. Students, in two groups, were helped in identifying species from different descriptions of certain characteristics (similar to plant identification keys).

NZ – What sort of activities did the students conduct in the woods?

***TC /SI* –** The students were involved first in a walk into the woods and then in two games, a pantomime and one related to finding trophic links. Operators (the teachers) tried to stimulate the curiosity and the spirit of observation of the students. The different species were pointed out during the walk by specifying their major and most evident differences, as well as the possible signs of the presence of animals (e.g. footprints). Furthermore, the operators highlighted the presence of biodiversity in the urban context and as an integral part of cities.

The pantomime game consisted in stimulating students to imitate animals occurring in forests. Students were asked to mimic both animals that can be found in more natural landscapes and those animals that can be found also in urban woodlands.

The second game regarded food webs and consisted in the identification of the links between different living beings of the forest. The main links were related to predation relationships. This enabled an explanation of the complex relationships between animals within forest ecosystems (Fig. 5.17).

Fig. 5.17 Padua Italy – The Woodland Close to School. Thomas Campagnaro

NZ – How was the outcome of the activities used – was the outcome of any of the activities incorporated into the larger Urban Forest management strategy; are the activities ongoing or passed to the next year's student body or...?

***TC /SI* –** A general report by the operators was prepared in which comments on the students reactions were reported. Students seemed to be very enthusiastic on all the activities. The project was part of a municipal program "Proposte per l'educazione alla sostenibilità A.S. 2018–2019" ("Proposals for sustainability education" School Year 2018–2019). The project has been confirmed also for this school year (2019–2020).

NZ – Is there a plan to follow any of the student's development to see what careers they select or if they remain connected to their spontaneous woods experiences as a result of their participation?

***TC /SI* –** No, as far as we know. It may be that this aspect could be introduced in the next years.

NZ – Was any information collected from the students to determine how they felt about there spontaneous woods experiences?

TC /SI – No information was collected on the experience from the students. This could be considered in the next years. I suspect that specific surveys and approaches should be used, especially considering the age of the children involved, and, therefore, may not be so straightforward if data should be then used to assess the experience.

British Columbia Canada – University of British Columbia Urban Forestry Program – Teaching Participation to the Urban Forester of the Future

In exploring varied and sundry Urban & Community Forestry participatory management initiatives, some elements critical to their success have crystalized. One of those elements is the importance of a working partnership between the official managers of the resource – the Urban Forester or their equivalent – and the participating community, regardless of whether the initiative is bottom up or top down.

I know of many professionals who do not see the value in citizen participation and avoid it if at all possible. From my own professional experience, I know it to be invaluable. So how do we break down these barriers? One way is during the educational process – how we teach our future Urban Foresters to create, work with and value would-be citizen stewards.

With this in mind, I decided to interview a colleague, Dr Cecil C. Konijnendijk, Professor (Urban Forestry), Program Director, Master of Urban Forestry Leadership and Project leader of the UFORIA (Urban Forestry Research In Action) lab at the University of British Columbia UBC, as follows:

NZ – As a Professor of Urban Forestry in a critically acclaimed program, how do you approach the participatory aspect of management? What is your thinking about this element of management and how do you project that?

CK – We do take this aspect very seriously, and integrate it in our courses from the very start of the Bachelor of Urban Forestry program. In our introduction course, UFOR 100 'Greening the City', one of four Modules deals with Urban Forest governance, and students e.g. learn about community engagement. In their final group project, they are asked to prepare a rudimentary plan for a local Urban Forest site, and they need to give due consideration to public participation. Later in the program, we work with more refined ways of including public participation.

I fully agree with you that Urban Forestry needs to be socially inclusive, and always make sure that this aspect is covered in my teaching. This is, for example, reflected in my own framing of the Urban Forestry concept, in which social inclusiveness is one of the key characteristics.

NZ – Do you find that most of your students are inclined to embrace the concept or are they somewhat resistant to the idea?

***CK* –** Most students in our program embrace the concept. We have a lot of socially and environmentally engaged students at UBC, and it comes natural for them to embrace participatory methods. These are also highlighted in other programs at our Faculty of Forestry (e.g. the Conservation degree). I think that perhaps the concept is more foreign to our transfer students from China, but after some time in the program, they also understand the importance of participation to Urban Forestry.

NZ – How have you been able to incorporate encouraging your students into the curriculum as well as sensitizing them to working with an interested citizenry in any of the following areas:

- reaching out to community leaders
- including community in planning meetings both as advisory and as equal voices;
- sharing knowledge in a digestible format to create informed stewards;
- working with community-based initiatives;
- encouraging experiential education in primary and secondary schools;
- supporting bottom up community-driven initiatives;
- other aspects – please share.

***CK* –** We cover participation aspects in many different ways. At a more basic level, students need to consider the interests of different stakeholders, including resident groups, in their group and individual assignments. In the more advanced courses, such as our Urban Forestry field camp and our final capstone course, students often do some active participatory work, e.g. by interviewing residents for their Urban Forestry projects. In both courses, we have held events with local stakeholders so that students can present and discuss their work. We often work closely together with entities such as the City of Vancouver for our project work, and stakeholder engagement is always high on the list. Moreover, through collaboration with the City, we share student project work with other stakeholders.

We encourage our students to volunteer and engage in environmental stewardship, as part of their overall development as well-rounded Urban Foresters. Some students have also tried to work with diverse cultures such as First Nations, but this community is very difficult to engage with in a more meaningful way, especially given the short time frames students have.

Students can also do individualized "directed" studies, and some have worked on topics such as school ground design. These projects typically have some form of community engagement component.

For larger, real-life participation and community involvement we technically need ethics approval from our university, which makes it sometimes difficult to go the extra step.

NZ – If your curriculum includes the writing of an Urban Forest Management Plan, do you require or suggest the inclusion of a community engagement/participation component in that Plan? Can you share examples of note that your students have included?

***CK* –** We work on aspects of Urban Forest management planning in several of our courses – but mostly in the 4th year. I mentioned the field course and capstone course. All of these plans need to have a community engagement component. It's a standard component of our approach to Urban Forest governance. Unfortunately I cannot share specific student works because of privacy limitations.

NZ – Does your curriculum require that the students create a participatory project or an educational offering?

***CK* –** We don't have such a requirement, but as mentioned we do encourage students to volunteer. Some of them also do public engagement and outreach work for cities, the Botanical Garden, or local NGOs through summer jobs or internships. We have the clear philosophy that Urban Forestry needs to be socially inclusive and promote public participation, and this is reflected in most of our Urban Forestry courses.

Community Forestry

... the very process of restoring the land to health
is the process through which we become attuned to Nature
and, through Nature, with ourselves.
Restoration forestry, therefore, is both the means and the end,
for as we learn how to restore the forest, we heal the forest,
and as we heal the forest, we heal ourselves.
– Chris Maser, *Forest Primeval*

One of the primary aspects of Urban Forestry is that the Urban Forest belongs to all of us – it is the people's forest. Just as with trees, our similarities as people – our genetics – are diversified not only by geography but by culture. If we think about the forest and then equate geography and culture with environment, we can see how geography and culture have enabled our genetics in oh so many ways.

In developing strategies for managing such a forest, it became clear that geographical and cultural differences would then equate with how a people's forest could and should be managed and thus Community Forestry had to be part of the larger concept of Urban Forest management. Just as a forest community is about much more than just trees, so Community Forestry is about much more than just the trees in the forest. It is about the relationship – the connection between a local people and their forested lands. It incorporates traditional practices and usage with local economy and land stewardship in a balanced and sustainable association. Its essence has been captured very well by the Center for People and Forests

"Community Forestry is a broad term used to describe models of forest management that give local people the majority say in making decisions. Similar terms include participatory forest management, collaborative forest management, social forestry, and community-based forest management. With an aim to reduce poverty, Community Forestry is participatory and should serve all community members equitably.

When we say local people, we mean indigenous people, ethnic minorities, individuals, and communities who have geographic, economic, social, and cultural relationships with local forest areas. Smallholder or household forestry, community or village management, and co-management arrangements within protected areas are all forms of Community Forestry.

The general aim of Community Forestry is to maintain healthy forests while serving local livelihood needs. The objectives of Community Forestry can vary: forest protection, household use, or commercial production. Multiple objectives often coexist because of the diverse needs of communities. These needs range from environmental (such as clean water production), to financial or subsistence-based (such as the use of forest products like timber and food), to spiritual and cultural." https://archive.recoftc.org/basic-page/what-community-forestry

RECOFTC – The Center for People and Forests has an extensive offering of training manuals and guides (in addition to other resources) which can be accessed through the following link:

https://www.recoftc.org/publications/browse?t%5B0%5D=87

The development of Community Forestry in practice has two distinct paths. In one approach, Government, as the holder of forested lands, gives communities use of the resource but the actual day-to-day management and the decisions that accompany such management remain in the hands of the State. The second approach and where we will focus our attention is when the State has legally delegated the management of the forest resource to the community. The greater the divestment of control and its transfer from State to local users, the greater the community empowerment and investment in sustaining the resource and the lower the management cost to the State.

Following is a sampling of how Community Forestry has been realized in geographically and culturally diverse communities around the globe.

The African Continent

Until you dig a hole,
you plant a tree,
you water it and make it survive,
you haven't done a thing.
You are just talking.
Wangari Maathai

When one thinks of Africa, one often thinks of civil unrest and the hardships that brings to human and environmental well-being. While there is unrest and all the difficulties that includes, there is also a well-spring of extraordinary initiatives across the continent and many of these initiatives fall within the Community Forestry framework.

Although I was not successful in obtaining as many interviews as I had hoped for African Community Forestry projects, they are so important and inspiring that I would like to share the information I have obtained in the hope that you, the reader, might also find them inspiring enough to stimulate further investigation.

The Green Belt Movement

A discussion of Africa-based Community Forestry would not be possible without presenting Dr. Wangaīr Maathai and her remarkable achievements, most notably the Green Belt Movement in Kenya. In 2005, after winning the Nobel Peace Prize I had the privilege of hearing Wangari lecture at Columbia University. Her passion for the Earth and all its children was palpable and the call to restore the earth permeated the very air of the auditorium. Wangaīr's presence was everywhere as if she herself was Mother Earth.

Throughout her distinguished career, she was primarily an educator in varied natural sciences, many of which were firsts for an African woman. Her early experiences with environmental and civic issues and movements, initially in the US where she studied, inspired her true calling – political and environmental activism. Through a great many trials and tribulations, Wangaīr persevered. Her observation that environmental degradation was a primary cause of Kenya's problems framed her pursuit of environmental restoration as a means to many positive ends. "Many women in rural areas said they were concerned about firewood, which was the main source of energy. They were concerned about water; there wasn't adequate clean drinking water. They were concerned about nutritious food, and they were concerned about poverty, especially among women. I immediately suggested that perhaps what we should do with these women is to plant trees. I saw the connection between land degradation and lack of water… Poverty is both a cause and a symptom of environmental degradation. You can't say you'll start to deal with just one. You're trapped. When you're in poverty, you're trapped because the poorer you become, the more you degrade the environment, and the more you degrade the environment, the poorer you become. So it's a matter of breaking the cycle" (2005 interview with The Progressive).

With the backing of the National Council of Women of Kenya (NCWK), the first of many Green Belt tree plantings took place on 5 June 1977 with the planting of 7 trees in a ceremony honoring female Kenyan environmental leaders. The Green Belt Movement was born with a handful of women gathering seeds and planting them. "Although I was a highly educated woman, it did not seem odd to me to work with my hands, often with my knees on the ground, alongside rural women. Some politicians and others in the 1980s and 1990s ridiculed me for doing so. But I had no problem with it, and the rural women both accepted and appreciated that I was working with them to improve their lives and the environment. After all, I was a child of the same soil. Education, if it means anything, should not take people away from land, but instill in them, even more, respect for it, because educated people are in a position to understand what is being lost. The future of the planet concerns all of us, and we should do what we can to protect it. As I told the foresters, and the women, you don't need a diploma to plant a tree."(Wangaĩr Muta Maathai – *Unbowed*, pp. 137–138.) Training moved from one village to many and 6000 tree nurseries were created. The apparent success of the initiative inspired Wangaĩr to expand the movement and, with the encouragement of the United Nations Environment Program, the Pan African Green Belt Network was established in 1986, enabling the sharing of methodologies and resources with 15 African countries including Tanzania, Ethiopia, Zimbabwe, Uganda and Rwanda. Many educational tools are published online and can be accessed at

https://www.pbs.org/independentlens/takingroot/classroom.html

MJUMITA – Tanzania's Sustainable Community Forestry Program

I learned of the existence of MJUMITA when I attended the first World Forum on Urban Forestry in November of 2018. Since then, I have explored the program in as much depth as possible and share my findings with you, the reader. http://mjumita.org/about-us/

The Community Forest Conservation Network of Tanzania, known in Tanzania as **M**tandao wa **J**amii wa **U**simamizi wa **Mi**situ **Ta**nzania – MJUMITA – came into existence in 2000. It grew out of a need for communities to share their community forest participatory management experiences and also to develop a dialogue with the Forestry and Beekeeping government agency – "a network of community groups involved in Participatory Forest Management (PFM) in Tanzania. The network provides a forum for capacity building, advocacy and communication for these groups" as well as "support to communities to demand and defend their forest rights where these are being threatened including in cases where private sector companies have attempted to take community forests."

MEDIC – Lagos Nigeria's Mental and Environmental Development Initiative for Children – From Beach Clean-Up to Tree Planting

I learned of MEDIC from an article written by Linus Unah for the the online Mongabay Newsletter https://news.mongabay.com/2019/10/beach-clean-ups-community-visits-and-compensation-to-fishers-build-environmental-awareness-in-nigeria/

MEDIC's website describes the initiative as "a non-governmental and pro-human organisation that aims at a total positive transformation and development of the child".

Aims & Objectives

To encourage children to be physically and mentally involved in the fight against an environmentally irresponsible Nigeria. To keep children responsible in their doings and enrich their self esteem.

To engrave in the minds of children the importance of integrity and honesty and deepen their sense of patriotism.

To make Nigeria a better place through children and to make the world a better place through Nigeria.

To create strong awareness on Global Warming and provide support in combating climate change through all relevant and legitimate means.

Our Vision

To create a cleaner, more habitable, more united world, where children are the purifiers and the spear-head of the revolution.

Doyinsola Ogunye, founder of MEDIC, has been working to raise awareness about plastic pollution and recycling since 2009. Her passion for the environment started after her family moved to Ajah neighborhood in 2002. She was astounded to find people living amid litter and cluttered waterways.

'Nobody was paying attention; it was just like a norm in that area,' recalled Ogunye, who studied to become a lawyer.

Trees originally dotted the coastline but they were removed as a result of spatial development and human pressure. While engaging students in a beach clean-up, Doyinsola has embraced the entirety of the beach environment and incorporated tree planting into the initiative. Linus connected me with Doyinsola and she has kindly shared information about her Tree Adoption Sustainability Plan:

NZ – What inspired the "Tree Adoption Sustainability Plan"? Is the project using any of the strategies developed by Kenya's Green Belt Movement?

DO – In 2016, my team and I planted over 500 coconut trees at what we now call the Kids Beach Garden. Unfortunately then we did not have an idea for the sustainability plan to protect the trees as the donations that came in for the project did not provide for this. I decided to encourage family and friends to adopt the plantings. I didn't even know about the Green Belt Movement when we started the Adoption plans.

NZ – The article states that you have 50 volunteers working on the tree-planting project in addition to 12 staff members. How were the volunteers recruited?

DO – Family and friends were of huge support in the beginning. We planted trees in schools , local communities, places of worship and parks. These were some of the major places we planted so we recruited from there as well. Social media announcements were also ways we recruited volunteers.

NZ – Are they from the Beach clean-up group? If so, how was this project presented to that group? If they are not from that group, how did you get the word out about this initiative?

DO **–** Some are from Beach Clean Up Groups, others from Corporate Organisations that volunteer with us**.** We sent messages , made calls and encouraged them to tell their loved ones as well . Most responded positively. We tried to go beyond the group mostly through social media.

NZ – Are the volunteers planting or is this something delegated to staff and the volunteers are adopting through funding of the project?

DO **–** Both applies . The staff planted trees through the new life project. Volunteers also come around during the rainy season.

NZ – If the volunteers are doing all of the above, was there training? If so, what did the training consist of and how long did it take?

DO **–** There has been no formal training. We keep teaching volunteers to plant trees practically on the spot to support a greener environment.

NZ – If it's staff that's performing all of the tasks, are they already knowledgeable professionals or are they community members that have been hired and have to receive training?

DO **–** Some hired staff receive training from Government parastatals that we work with.

NZ – Is the project incorporating any local indigenous practices into the performing of some or all of the tasks?

DO **–** Yes. For example, to keep goats away from the trees, we mix their feces with water and sprinkle it on the trees' leaves and bark. This will prevent them from eating the tree.

NZ – Who is supplying the Coconut Palms that are being planted? Are they being grown especially for this project or…? Are they being donated or is there a subsidized price or…?

DO **–** All applies

NZ – Is there an establishment protocol such as watering? If so, for how long and is that undertaken by volunteers or staff?

DO – Both volunteers and staff water the plants 3 times a week

NZ **–** Have you given any thought to establishing a nursery where the trees that are being planted could be started? This could also expand to provide trees for other Community Forestry initiatives.

DO **–** Yes we have tree nurseries in 3 villages now.

NZ – Have you been able to introduce some of the environmental information driving the program into the schools and have it incorporated into school curriculum?

DO – Yes, we have been able to introduce students to what we are doing and why via excursions

NZ – Have any steps been taken to gain protection for the areas that have been planted which would also protect this area from future development?

DO – Yes, this is very important. We are starting negotiations with the Federal Government this year.

Bhutan – Protecting and Sustainably Using Community/ Social Forests

The Asian kingdom of Bhutan is located between India, from the south to the east, China to the north and Nepal to the west, in a rugged mountainous area of the Eastern Himalaya with a range of elevation from 200 m to more than 7000 m. An estimated 70% of its land-locked 38,394 km^2 is covered by forests, divided between sub-alpine coniferous forest at higher regions and broadleaf deciduous forests at lower elevations.

After instituting the Forest Act of 1969, which nationalized all the forests, the government came to realize that "sustainable forest management could only be achieved with the local people: People's participation is key to conservation and the utilization of forest resources", stated a royal decree in 1979. "This recognition prepared the ground for a step-wise, albeit slow, paradigm shift towards a more decentralized and people-centered approach to forestry, and the government started developing approaches to involve local people in forest management again. One of these approaches is Community Forestry." (Asia Brief, East Asia Division. June 2013).

Early 2013 saw the 500th national forest turned back to the people with one fourth of Bhutan's entire rural population becoming members of a CFMG – a Community Forestry Management Group. These groups are responsible for planning, managing, monitoring, and improving their community forests (CFs). In some districts, more than half of all rural households participate. This illustrates that Community Forestry has effectively become a broad-based movement and without doubt, the most inclusive community-based civil society enterprise in the country. "The establishment of CFs is now demand-led and fuelled by the high interest of rural communities to take over responsibility for the management of their surrounding forests." (Asia Brief, East Asia Division. June 2013).

An important partner in the development and implementation of the CFMG approach to the management of the national forest resource is the Swiss organization Helvetas. With initial support from Mr. Chado Tshering, head of the Social Forestry Division (SFD), Dr. Kaspar Schmidt, as the Helvetas project advisor of the

Participatory Forest Management Project (PFMP), provided overall guidance and direction for the strategic framework. Kaspar was born in Zürich Switzerland. He received his MSc in Forest Science from ETH Zürich and his PhD from Reading University (UK) where he did research on the people and forests in Southern Kyrgyzstan. He has continued his work on international forestry, natural resource management, climate change and development through his affiliation with the Swiss independent NGO Helvetas. "From 2009 till 2013, I worked as technical advisor of the PFMP in Bhutan implemented by the Royal Government of Bhutan and Helvetas and funded by the Swiss Agency for Development and Cooperation SDC."

Kaspar has kindly shared his Bhutan project experiences, as follows:

> **NZ –** How was the elaboration of the National CF Strategy which took place in April of 2009 participatory? Were community leaders included in the process and, if so, how were they included? How much value did their contributions have – purely advisory in the "taken under consideration" category or well-considered voices which were included in the outcome?

KS – For all the driving forces behind the elaboration of the National CF Strategy, i.e. the Social Forestry and Extension Division of the Ministry of Agriculture and Forests, Helvetas and Dr Don Gilmour, the consultant who led the process and drafted the strategy, it was very important to have a participatory process in which the Community Forest Management Groups (CFMGs), in particular as well as their leaders would have an important say. Experienced members of the Executive Committees of selected CFMGS from all parts of the country actively participated in the two-day workshop held in mid-April 2019 in the capital city Thimphu. In his role as workshop facilitator, Don Gilmour, a very experienced pioneer of CF in the Himalayas, made special efforts to ensure that community members had a voice at the workshop and throughout the process of elaborating the strategy. The project provided interpretation for their inputs to working groups and their presentations at the workshop. After the workshop, Don Gilmour went on a fieldtrip to Southern and Central Bhutan to meet with a range of different stakeholders. On his trip, Don and the Bhutanese colleagues accompanying him visited five Community Forests and met with members of their CFMGs.

In a context with a relatively hierarchical society and an often dominant government and administration, one needs to amplify the voice of communities to make them heard. Looking back at the process, I think that overall we managed to do this quite well. Growing confidence and interest amongst communities to get involved in forest management, on the one hand, and support from "CF champions" in key positions in the administration including the then Minister of Agriculture and Forests who were themselves convinced of the benefits of the involvement of local people in forest management and conservation, on the other hand, helped us to 'advance the cause'.

> **NZ –** The term *sustainable* permeates a great deal of the information as part of the governing criteria with participating communities. How is the need to access the

forest's tangible resources, e.g. harvesting, combined with the need to do that sustainably?

KS – Bhutan has a long-standing tradition of nature and forest conservation and takes great pride in this. Of course, there is forest degradation etc. occurring, but generally CFMGs and foresters alike are very conservative when it comes to harvesting.

The process to establish a Community Forest includes various participatory exercises to collect and map the knowledge held by the local community about their forests and other resources, a participatory forest inventory and assessment of non-timber forest products (NTFP) available. The inventory and the NTFP-assessments are conducted by community members with technical support of forestry extension agents. Based on the inventory results, sustainable harvest levels for timber and fuel wood are calculated. Guidelines on a range of NTFPs determine sustainable harvesting levels for the NTFPs. The whole process to apply for and set up a CF and write a CF management plan is described in great detail in a CF manual. Forestry extension agents are trained in the process and use of the manual, technical guidelines etc. to support each community in this process.

As we know from all over the world, local communities are very effective managers ensuring conservation and sustainability of their forest once entrusted with long-term tenure rights for forests. This has also proven to be true in Bhutan. There have been a few cases of allegations of misuse and unsustainable harvests in CFs, some of which a Bhutanese colleague and I have been asked to look at in greater detail and analyze, which we did and realized that in each of these cases, the allegations were unsubstantiated.

In the Bhutanese system, CFMGs get forest management rights based on a forest management plan that is valid for 10 years. In case a CFMG is found to violate its responsibilities and not to adhere to the plan, the government reserves the right to cancel the forest management plan and take the resources back under its direct control and management.

NZ – Is there an educational component which offers guidance and provides essential information on managing forest ecosystems sustainably? Is this a prerequisite for CFMG membership? Who provides the training and is it free of charge?

KS – Yes, communities and individual farmers are supported and advised by forestry extension agents in CF, the establishment of a CF and other forms of participatory forest management. These forestry extension agents are graduates of the College of Natural Resources (CNR) that is part of the Royal University of Bhutan. They provide decentralized advisory services to the community and individuals in the districts and communes of Bhutan and play a key role in raising awareness about sustainable forestry, CF etc., facilitating the process of CF establishment and accompanying the implementation of CF management plans with technical advice and guidance. Rural extension services (forestry, livestock, agriculture) are funded by the Royal Government of Bhutan and provided at no cost to its beneficiaries.

In the CFMGs, members themselves organize trainings in forestry, typically with support of forestry extension agents. The PFMP, the project I worked on, also supported the development of peer learning approaches for the CFMGs to exchange experience and learn from one another (CF study tours to other districts, joint trainings etc.) and promoted the establishment of networks of CFMGs. During its time, the PFMP also organized trainings on various topics (e.g. silviculture, logging, NFTP management and harvesting, group governance). Hereby the project usually applied a "train the trainer" approach, i.e. the project trained trainers, for example interested, previously trained community members or forestry extension agents who then trained CFMG members.

> **NZ –** I have noted that the collective encompasses different ethnic backgrounds. Can you give examples and indicate how traditional indigenous Community Forestry management practices might be accommodated?

KS – Yes, the different ethnic groups living in Bhutan all have their own traditions. Depending on their culture, religion, beliefs, resources available, they manage forests and other natural resources differently. I remember visiting a CFMG in Central Bhutan. When talking about their forests, they mention that, for religious reasons, they were banned from entering and harvesting bamboo from their forests during certain periods. It turned out that the ban coincided with a period that was important for the regeneration of bamboo and other key species of this forest.

> **NZ –** Who is providing all the necessary tools and equipment for timber harvesting? Is it being provided free of charge or are the communities required to purchase /rent?

KS – Usually some individuals in the communities have the necessary tools and equipment for timber harvesting. The PFMP has provided safety equipment for free (helmets), but not hardware such as chainsaws. CFMGs can themselves decide on how to invest the community funds they generate from marketing forest products. Thus a CFMG with productive forests can also make an investment into procuring tools and equipment.

> **NZ –** Are private sector companies still involved in the process, e.g. timber sales and does that impose any difficulties in the sustainable aspect of management as all too often, the private sector is motivated solely by profit and sustainability of the resource and its habitat pay the price?

KS – Yes, private sector actors are involved in value chains of timber, NTFPs and other forest products. I have met several private sector representatives who were committed to sustaining the natural resource base of their business and interested to team up with CFMGs to join forces for harvesting of timber or NTFPs and subsequent processing and marketing steps. Obviously, there are people with other motives, also in Bhutan.

As with many other things, timber sales are highly regulated in Bhutan. It took the project a while to develop marketing guidelines for timber harvested from CFMGs and facilitate the adaptation of the then existing system of permits for the transport and sale of timber by CFMGs.

NZ – Have capacity issues been addressed so that capacity keeps up with demand?

KS – Capacity development by members of CFMGs, members of the Executive Committee, forestry extension agents, researchers, other foresters and other stakeholders has been an important dimension of the PFMP from phase I until phase III, but with different foci and methods. At the beginning, it was key to build awareness about CF and stimulate interest in the approach. As the number of groups was limited for a long time, the project could apply a direct training approach in the beginning and then change to the above-described "train the trainer" approach with the growing number of groups and interest in CF. Training committee members is key to ensure sustainable management of forests, but also good group governance. Members of the Executive Committee play a crucial role in both aspects. Many CFMGs have a few Executive Committee members who act as forest watchers. In other groups, CFMG members take turns in checking their forests. These are answerable to the Executive Committee of the CFMG and the committee again to the general assembly of the CFMG.

Guatemala: Concepción Chiquirichapa

A town located in southwestern Guatemala's Quezaltenango Department, is home to approximately 18,000 people, mostly of indigenous Maya Mam origin. It is also home to Siete Orejas, the volcanic mountain that is sacred to these Maya Mam peoples.

Fig. 5.18 Volcán Siete Orejas By Simon Burchell – Own work, CC BY-SA 3.0, https://commons.wikimedia.org/w/index.php?curid=12439508

The slopes of Siete Orejas (Fig. 5.18) were originally cloud forest but, following an eruption, the slopes were dominated by successional grasslands which were being managed for grazing sheep. Economically, Concepción relied on potato farming, thanks to ideal growing conditions and a soil enriched with organic leaf litter. Concerned for their water supply, the Concepción community decided 40 years ago to restore the cloud forest using their traditional indigenous forestry practices.

"For the Maya Mam, water springs are very important, because they are the source of all life," says Martha Tax, head of operations in the region for the Swiss NGO Helvetas, which has been conducting forest conservation, drinking water improvement and economic development programs in Concepción and other local towns since 2006. "They consider them as sacred places and pay respect to each one of them."

This reverence is on display in Concepción, where the lids of the springs bear vases of fresh flowers that town residents replenish each day. And it extends up the mountain to the springs' sources. Rumualdo López, a spiritual guide and member of Concepción's municipal council, says the 22 altars on the mountain are located at "energetic portals" used by Mother Nature to give birth to the clouds.

"We know that these particular places are the ones where the mountain itself transforms into water," he tells Mongabay. "That is something our grandparents told us." (quoted from the Jorge Rodriquez three-part profile of the Concepción Chiquirichapa community and its stewardship of its sacred cloud forest, published by Mongabay.org in its series: Indigenous Peoples and Conservation).

With the sheep gone, forest and wildlife began to return to the high peak area and the community began to take notice. Beginning in the 1970s, sixteen community park rangers began planting trees, restoring the cloud forest. Today, two of those original park rangers, Oscar López and Francisco Escalante, along with six more rangers from the Department of Protected Areas (DAP in Spanish) are in charge of "…making rounds, planting new trees, cleaning trails and taking care that no one destroys the forest or harms the wildlife."

Following is Concepción's community forestry journey In their own words with text excerpted from Jorge Rodriquez' investigative offering and quotes from Marcelino Aguilar, Oscar López and Francisco Escalante.

First steps toward conservation – With no technical background or contemporary concept of conservation, Concepción's people relied on the ancient Maya knowledge of their grandparents to begin restoring Siete Orejas' forest in the 1970s. Concepción's municipal government turned to the local Indigenous Council, a group of Mayan leaders appointed by the community, to decide how to proceed. According to López and Aguilar, they decided to limit firewood collection and give several people responsibility for caring for the forest, including guarding against firewood collection, planting new trees and maintaining trails. Initially these community park rangers were appointed annually and worked for no pay.

"We worked for free since the idea of preserving the mountain began," Escalante says, as he wields a machete to slash firefighting trails on Siete Orejas. "For us it was a great honor to be considered by the people to take care of the forest and it is something that we continue doing with pride."

"The results of their work are there for all to see in the form of Siete Orejas's forested slopes, clean and more abundant water, and thriving potato fields.

The flourishing forest is a small, green flicker of hope in Guatemala's western highlands, a region beset by poverty and sky-high rates of malnutrition among children."

If we look at the entire evolution of this example of community forestry, we can see how it grew and what that growth has accomplished:

Protection for the forest:

As a result of the creation of Guatemala's National Council for Protected Areas (CONAP), the 1990s saw a "protected area fever" awakening in everyone who owned natural forest. "Local governments, private individuals and the Guatemalan government sought the declaration of their forests as protected areas."

- In 2008, Concepción designated "a 1,200-hectare (4.6-square-mile) park on Siete Orejas named Kum Kum Wutz. Under Guatemala's Protected Area Law, it is classified as a regional municipal park, which means that people can harvest certain natural resources, including wood, organic matter for fertilizer and medicinal plants, as long as they do not harm the forest." "The forest provides mushrooms and other kinds of food, as well as medicinal plants for the people to use. If there is a tree that fell down of natural causes, we allow the people to get in and chop it to get firewood," Aguilar says.
- Concepción also designated five other small forests, totaling 25 hectares (62 acres) scattered around town, as protected areas. The DAP got equipment and infrastructure and began organizing programs to sustainably generate income from the forests, such as the sale of leaf litter and a way of tracking the value of local people's volunteer forest maintenance work.
- It also enrolled 190 hectares (470 acres) of Kum Kum Wutz and all five of the small forests in a program through the National Institute of Forests (INAB by its Spanish initials) that pays landowners to commit to preserving forest cover on their land for a period of 10 years. This has generated $22,000 for the DAP's operations, according to Aguilar, and the office plans to enroll more land in the coming years.

Education:

- Between 2004 and 2009, Helvetas, an independent Swiss development NGO, worked with Concepción and several other communities, to educate community park rangers and advise municipal staff, like Aguilar, on creating and maintaining protected areas. This was key for the creation of the DAP, as part of the Environmental Commission, in 2004.
- With the help of international donors, the DAP has also created a network of socio-forestry trainers who help educate people about environmental topics. This aid also allowed the DAP to install a meteorological center on top of the mountain, with the objectives of measuring the effects of climate change, creating policies to adapt to it and generally becoming a more resilient town. Other projects have focused on training people in soil conservation techniques and promoting the diversification of crops beyond potatoes (Fig. 5.19).

Fig. 5.19 An educational sign in Kum Kum Wutz park encourages people to plant trees, Jorge Rodríguez, Mongabay.org

Helvetas has also been involved in re-teaching people about medicinal plants that grow in the forest to help them reconnect with their cultural roots and save some money on health care.

Sustaining reforestation efforts:

- In 2015 the DAP under Aguilar, with technical assistance from Helvetas and financial aid from the United States Agency for International Development (USAID) set up a nursery to grow trees for the reforestation efforts. In 2018 it planted more than 15,000 new trees; it plans to double that number for 2019. Concepción is now considering joining an INAB incentive program that encourages landowners to supply seeds to other parts of the country for their own reforestation efforts. The town's pine, oak and cedar trees, as well as others like aguacatillo (*Persea caerulea*), are now old enough to produce seeds in quantity.
- The local restoration of the pinabete, or Guatemalan fir (*Abies guatemalaensis*), is one of the Concepción conservation program's chief successes. The tree is a protected endemic species that the IUCN classifies as endangered due to the illegal traffic of its branches for use as Christmas trees by Guatemalan families. To date, Concepción has restored about 7000 pinabete trees, the equivalent of 7 hectares (17 acres). "Through talks with the villagers, we have managed to eradicate the illegal commercialization of this tree," Aguilar says.

Putting the public in the public forest assets:

- Aguilar is keenly aware that the community's participation over the past four decades has been key to the successful restoration of Concepción's forests, and he has taken pains to cultivate it.
- In 2016 the DAP started the Environmental Services Compensation Program as a way to involve the entire community in caring for the protected areas. With the help of local decision-making bodies called Community Councils on Urban and Rural Development, the DAP invites people to donate a day of work, valued at $7.73, as a sort of payment for what the forest provides them in water, organic matter, natural medicines or other services. During the first year, 60 people carried out tasks such as maintenance, cleaning and reforestation, which translated into a community investment of $463. In 2017 the number grew to 300 people, for an investment of $2319.
- "Through this program, people are realizing how important it is to take care of the forest, because it is thanks to it that we have water and a lot of other benefits," Aguilar says.
- One nascent DAP effort to bring the public into forestry and benefit the wider ecosystem is the creation of so-called energetic forests. In 2005 it became clear that people were extracting too much leaf litter from the protected areas for use as fertilizer on the potato fields, and it was negatively affecting the forest. So the DAP decided to charge 13 U.S. cents per bag of leaf litter, place a limit of 25 bags per person per week, and maintain a record of who and how many bags people extract. To take further pressure off the protected areas, in 2016 the DAP, with funding from international donors, began encouraging private landowners to set aside forest as a source of leaf litter they could sell. It has the added benefit of offering landowners an incentive to keep their forest standing, although so far only one has signed up.
- The cumulative effect of all this outreach and education has been a clear uptick in people's involvement in conservation activities like planting trees and picking up trash in the forest. A yardstick of this is increased participation by women, even though they tend to have a secondary role in Concepción's public life. "Before, they were the first ones who came to complain when there was a shortage of water. But now, through environmental education programs, they learn where water comes from and why it is important to protect our natural resources," Aguilar says.
- The DAP's programs are having an effect on local attitudes, observers concur. "People are learning and their attitude toward forests and wildlife in general is very different from what it was a few years ago," says Martha Tax, head of operations in the region for the Swiss NGO Helvetas.

Eco and cultural tourism:

- The national tourism authorities are working with Aguilar's office and members of the municipal council to promote tourism in the area. In 2015 the Guatemalan Institute of Tourism (INGUAT, by its Spanish initials) added Kum Kum Wutz to its list of destinations in the region. The park has a 650-meter (2132–foot) zip line, camping zones and a great view, and its municipal tour guides have been trained in first aid, rescue maneuvers and customer service. The park has also been included on a trekking circuit that takes in natural destinations in the western highlands of Quetzaltenango.

- Part of the rationale for focusing on tourism is that the field offers an appealing way of life for young people, one that Aguilar and others hope can inspire them to stay rather than migrate to Guatemala City or the United States, as many do to escape the poverty and poor prospects at home. "For the new generations, the old spiritual practices are boring and uninteresting. But tourism allows them to meet people from different parts of the world, get involved in exciting activities and it gives them the opportunity to see what nature has to offer," Tax says.
- Officials are also considering how Concepción's Mayan heritage might attract visitors. "We have made some trips into the mountain with people who are not spiritual guides. They are interested in learning about our spiritual connection with Mother Nature and want to help change the path in which humanity is living right now," says Rumualdo López, the spiritual leader and municipal council member.

Many thanks to Jorge Rodriguez, a Guatemalan photographer and journalist, for his intimate, animated reporting and to Mongabay.org for presenting this inspiring contribution to the global community forestry initiative.

https://news.mongabay.com/2019/02/how-a-maya-town-restored-its-sacred-cloud-forest-and-water-supply/

The creation of a thousand forests is in one acorn.
– Ralph Waldo Emerson

Participation Through the Intangibles – Social Activism as Art

We can see from the experience of Odin
that the image of the tree was the template
within which all of the sacred world could be apprehended.
The tree was the framework within which one "flew" to these Otherworlds.
And since the exploration of sacred space
was also a quest into the nature of human consciousness,
the tree was regarded as an image
of the ways in which we, humans, are constructed psychically.
It was a natural model for our deepest wisdom, our highest aspirations.
Brian Bates, Sacred Trees

There are many ways in which we can participate, either directly as enumerated above or indirectly. In Chap. 1, we discussed the extensive array of intangibles celebrating the tree and the varied landscapes the tree has populated over these many millennia. Looking at today's world in comparison with the world of our origins, it is clear to me that intangible inspiration is of greater importance than ever. The turn of a phrase, the sound of a chord, the stroke of a brush can open us up to the vast world of possibilities. The intangibles – the craft of creating visual art, sound art and the spoken and written word – offer profound mechanisms for inspiration, an essential ingredient in the "taking ownership" recipe.

This final participatory component is a visual offering by artists I am personally connected to – a celebration of their connection to trees that inspires, invites, informs.

"Watercolor painting is my way of studying patterns in nature, sharing peace in the strokes of tree branches and bird feathers. Bark is not brown, it is a patchwork of soft blues, reds, and amber. By articulating the delicate details, I hope to evoke wonder and compassion for trees and wildlife and inspire a collective approach toward coexistence.

In autumn, the Western larch, *Larix occidentalis,* is a bright golden gem in a sea of evergreen. One of the world's few, the larch is a deciduous conifer that looses its needles in winter" (Fig. 5.20).

Erin Hauer

Fig. 5.20 Western Larch. Erin Hauer

"An der Hochschule für Gestaltung und Kunst waren wir zum Zeichnen in der Natur. Unterhalb der Museggmauer entdeckte ich einen alten Platanenbaum und war sofort fasziniert von der seltsamen Stammstruktur. Da besuchte ich diesen während drei Wochen fast täglich"

[At the College of Art and Design, we went out into nature to draw. Below the Musegg wall I discovered an old plane tree and was immediately fascinated by the strange trunk structure. I visited it almost daily for three weeks…] Margarit Eberle-Jörg (Fig. 5.21).

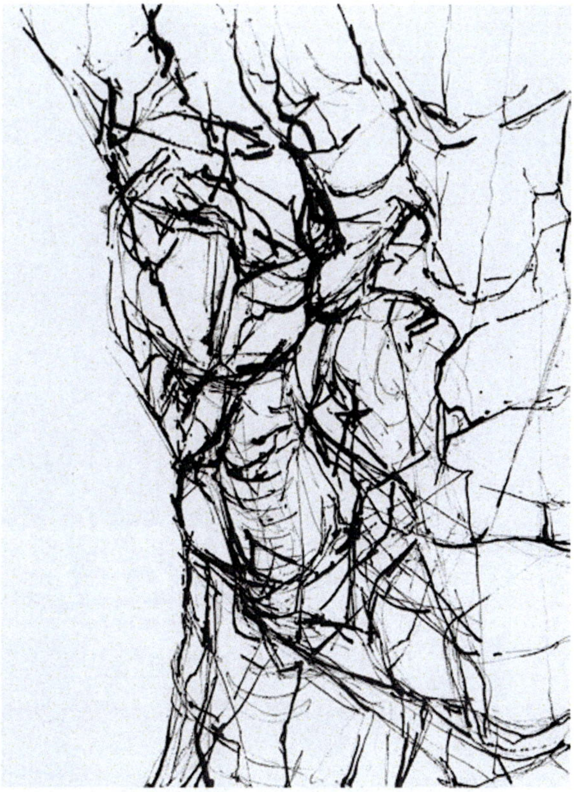

Fig. 5.21 Platanus study. Margarit Eberle-Jörg

"Being from a rural area, I always look for the green and the living when I visit a city. Parks and public gardens are my place to go. While exploring the topic of Urban Forestry and Urban Gardening for a school project, I realized that my feel good places are affecting a majority of a city's health. Not only do they reduce the greenhouse gases and city's heat, they also stem traffic noise and ultimately optimize people's quality of life. I needed to share the importance of the Urban Forest with a wider range of people, so I created this moving poster in the hope that its liveliness and energy invite the viewers to consider my version of a utopian world." Kathrin Blümli (Fig. 5.22).

Fig. 5.22 Urban Forestry concept poster. Kathrin Blümli

"Das Rauschen der Blätter im Wind, die Taktilität von Baumrinden oder Schatten spendende Ort; Bäume sind für mich an sinnliche Erinnerungen gekoppelt. In dieser Fotografie faszinierten mich die natürlich, geschwungenen Linien der Baumstämme und die undurchschaubare Gesetzmässigkeit der Natur. Meine Intention war es, durch meinen Eingriff dies zu verstärken und bewusst zu machen."

[The rustling of leaves in the wind, the tactility of tree barks or shady places; trees are for me linked to sensual memories. In this photograph I was fascinated by the natural, curved lines of the tree trunks and the inscrutable laws of nature. My intention was to intensify and make conscious this through my intervention.] Barbara Davi (Fig. 5.23).

Fig. 5.23 Wald collage. Barbara Davi

Portraits of trees
City – Forest – City forest

"This work examines different habitats in which trees can grow and live. I have created this series of photos to document different environments in which trees exist. For this purpose, I have chosen trees in urban areas, the city forest, and the Swiss National Park in Graubünden. These three spaces showed me very contrasting versions of the environments in which trees exist. The three comparison locations were chosen deliberately, because for me, they also deal with the question of where we "allow" nature in our environment. Through the exemplary view of the three habitats, the idea arises that nowadays, we could regard "nature" as man-made or man-determined. In all three examples, it is the human being who decides if and how nature looks. In one case he leaves it to itself to be wild. In the city, every tree is a part of urban planning considerations, and in the city forest, forestry on the one hand and recreational areas for humans on the other hand decide about trees.

The project was an enriching experience for me, as I spent a lot of time in the forest and, by consciously focusing on the differences, I became more aware of the forest. If you look closely and pay attention to the trees, their surroundings and the soil, you will discover many small details that point to human intervention in nature. The contrast between the city forest in Zürich and the National Park was striking and very impressive. The work made me think a lot about nature and I wonder how it could be possible for man and nature could live in a more balanced relationship." Valentin Egli, 25.5.2020 The Tree & Me - Seminar ömomedien – ökodata – ökoaestethik. Hochschule für Gestaltung und Kunst, Basel (Fig. 5.24).

"Plan B. Trees as partners for a climate-friendly urban future" exhibition poster

"Das Konzept für „Plan B. Bäume als Partner für eine klimafreundliche urbane Zukunft" entspringt meiner überzeugung, dass das es bei der Vermittlung von Umweltthemen Kunst und Design braucht. Denn nur durch sie werden Umweltprobleme, also Disharmonien zwischen menschlicher und nicht-menschlicher Natur, über die reine Verstandesebene hinaus erfahrbar. Das ist ein zentrales Moment, wenn wir in einer ökologisch nachhaltigeren Welt leben möchten. Die Ausstellung bringt interdisziplinär Kunst und Gestaltung mit naturkundlichem Wissen zusammen. Im Mittelpunkt der Ausstellung steht die Erkenntnis, dass Bäume wichtige Partner sind für unser Leben in der Stadt und dass ihr Dasein vor dem Hintergrund des Klimawandels noch essenzieller für uns wird."

[The concept for "Plan B. Trees as partners for a climate-friendly urban future" stems from my conviction that art and design are needed to communicate environmental issues. It is only through them that environmental problems, i.e. disharmonies between human and non-human nature, can be experienced beyond the purely intellectual level. This is a central moment if we want to live in a more ecologically sustainable world. The exhibition brings together interdisciplinary art and design with "nature" knowledge. At the heart of the exhibition is the recognition that trees are important partners for our life in the city and that their existence is becoming even more essential for us against the background of climate change.] Julia Sommerfeld. (Fig. 5.25).

Fig. 5.24 Baumportrait. Valentin Egli

Fig. 5.25 Plan B: Bäume als Partner für eine klimafreundliche urbane Zukunft Exhibition Poster. Julia Sommerfeld

References

Sitzia T, Campagnaro T, Weir RG (2016) Urban Ecosyst 19:475. https://doi.org/10.1007/s11252-015-0475-3

Feng Y, Negron-Juarez RI, Patricola CM, Collins WD, Uriarte M, Hall JS, Clinton N, Chambers JQ (2018) Rapid remote sensing assessment of impacts from Hurricane María on forests of Puerto Rico PEERJ Pre-prints.

Chapter 6
Making a Difference – Questions Asked and Questions Answered

What can we learn from these varied model approaches and their creative applications; what benefits have been realized for trees and people alike; how can we continue to enhance these endeavors?

Keywords Actionable participation · Continuity · Empowerment · Growing · Investment · Potential · Taking ownership · Taking responsibility · The teachable moment

The care of the Earth
is our most ancient and most worthy,
and after all our most pleasing responsibility.
To cherish what remains of it
and to foster its renewal
is our only hope.
Wendell Berry

While this tree homage is a celebration of programs and endeavors which recognize that trees are worthy recipients of our efforts on their behalf, it also presented a critically important opportunity to look at a sampling of the breadth and depth of what people all over the world are doing for their trees and, as a tree professional, add my own experience-based thoughts to the mix.

While the Urban and Community Forest holds many tangible and intangible benefits, actionable participation brings another reward – being empowered to take ownership. Taking ownership as part of a collective – one's community – can be equated with taking responsibility as an individual for something that benefits everyone. As we collectively grow our respective Urban and Community Forests, we grow ourselves and, in so doing, we grow our society.

If one were to try to devise a recipe for involving citizens in Urban and Community Forest management, my own experience with such endeavors recognizes several ingredients – two of which are caring and interest. If you care and you are interested, most things are possible. That's as true for the tree professional as it is for the lay citizen. For the tree professional – is it just a job or do you truly care about the well-being of the trees you are responsible for and the citizens they are asked to coexist with? For the citizen, do you take notice of the trees that are a part of your community and, if they look as if they are struggling to survive, have you thought of

N. Zürcher, *Connecting Trees with People*, Future City 16,
https://doi.org/10.1007/978-3-030-94534-3_6

how you might improve their situation? Have you asked yourself the *What can I do* question?

In addition, the many decades I had spent developing and implementing such programs brought with it a genuine evolution in terms of what is possible. At first, the possibilities seemed very limited. Visions of Citizen Pruners inflicting their minimally trained aesthetic on urban trees gave me great pause. Rather than throwing my hands up in the air and abandoning the training, I started to vary my approach to the *WHAT* that I was sharing. I found that, as I tested the waters, it became more and more clear that each situation combined with my own flexibility should guide what responsibilities could be offered, how they could be shared and what the anticipated outcome might be.

Do not be afraid to go out on a limb …
That's where the fruit is.
– Anonymous

What I learned over these many years is never underestimate potential. If a citizen cares and shows an interest, that's fertile ground and deserving of an investment of time and knowledge. If the tree professional cares and takes an interest, the time it takes to offer, in a genial manner, meaningful, accessible knowledge is a worthwhile investment. Taking advantage of the teachable moment – a willingness to answer questions and share what you know in an accessible language – will be rewarded many times over.

A number of the questions posed during all the varied interviews had a similar thread – education, continuity and participant diversity. Rather than dissect the responses specifically, I would offer you, the reader, an overview of my observations and impressions.

Education: Experiential learning opportunities and the effective distribution of stakeholder-specific accessible information is critical to the successful planning, management, protection and preservation of the Urban and Community Forest resource. An informed steward is an invaluable asset who can:

- successfully perform critical tasks, e.g. mulching open soil or watering during drought. These are tasks that are time consuming and usually fall by the wayside but are essential, not just for the tree, but for the landscape into which the tree has been planted. Landscape equates with soil and soil that cannot function will not grow a healthy tree.
- facilitate long term planning. If a community participates in deciding where trees will be planted in their neighborhood, that's a first step in the process of taking ownership. If community participants know what a tree actually requires in order to grow and how that benefits them, they will have a better acceptance of steps that need to be taken to accommodate trees and also a greater tolerance for things that the uninformed find an annoyance, e.g. leaf litter. If trees' needs are accommodated both below and above ground, there will be less damage to surrounding infrastructure and to the tree trying desperately to create adequate growing space – an important selling point for the planning of tree-friendly hardscape.

• be an exceptional advocate, especially when it comes to funding Urban and Community Forest management. It's one thing for a decision maker to hear from a constituent who likes trees. It's quite another when that citizen is conversant enough to make valid suggestions or challenge the existing state-of-affairs. Enough of those voices, raised in a timely manner, can facilitate very positive outcomes – improving an existing management plan or creating the impetus to develop and implement a new one.

• once trained, be a trainer of others in the community – a simple train-the-trainer approach to expanding informed eyes and ears. Such strategies plant the seeds of community empowerment and foster a sense of place-making and well-being. By taking care of our community's trees, we are taking care of each other as well.

That land is a community is the basic concept of ecology,
but that land is to be loved and respected is an extension of ethics...
We abuse land because we regard it as a commodity belonging to us.
When we see land as a community to which we belong,
we may begin to use it with love and respect.
Aldo Leopold -

Continuity: Once engaged and trained, it would be an unfortunate loss of a valuable resource if participating citizens were not encouraged to do more. Regardless of the tree-related initiative, from the planning and preparing for and the selecting of a tree to be planted or assessing a tree that is already in the ground to the actual planting of that tree, participating in the *what* that comes next is a critical part of the process. Even the isolated tree provides habitat for an extraordinary array of diversity from microbes and invertebrates to birds and other urban wildlife. That habitat can only be supported and enhanced if the tree can grow. If one understands that in planting a tree, you begin an entire universe, why not offer the possibility of stewarding that universe or encourage and assist those citizens who step forward and offer helping hands to do just that. If citizens care and they are interested, they could be and should be incorporated into stewarding the possibility of longevity.

The true meaning of life is to plant trees,
under whose shade you do not expect to sit.
– Nelson Henderson
Northern Manitoba CA farmer

Participant diversity: Since the Urban Forest belongs to all of us, it then stands to reason that participation opportunities should be as diverse as the healthy ecosystem we are trying to realize combined with the needs of all community members, trees and people alike. It's the disenfranchised communities that all-too-often have the fewest opportunities to participate. These are the communities that receive the brunt of polluting installations but have the lowest percentage of public green spaces to help mitigate the pollution. They are also the communities who have a limited exposure to diverse job opportunities. Creating programs that engage with these communities and introduce them to the wonderful world of *green* – careers as well as place-making, is as empowering for the individual as it is for the community.

Creating opportunities for any and all aspects of participation must be made available to all citizens. As populations become more and more diverse, Urban and Community Forestry initiatives in the form of community gardens, urban allotments or other urban or rural agricultural-oriented programs offer additional opportunities for participation while enhancing an under-served community's access to nutritious food or economic growth. These programs must be accommodated within our publicly held green spaces and undertaken with the community's direct involvement in the planning, design and development processes. Such involvement supports the implementation of social and environmental justice concepts and usually results in empowering a sense of ownership and thus the protection and preservation of the site. The development of such urban spaces may require a bit of handholding on the part of Management personnel but thoughtful and culturally sensitive input will often facilitate an informed, pro-active community involvement in site use and maintenance.

May we open to a deeper understanding
and a genuine love and caring
for the multitude of faces
who are none other than ourself.
Sensai Wendy Egyoku Nakao
Abbot Emeritus, Zen Center of Los Angeles

To further expand on the human diversity perspective of Urban and Community Forestry participatory management, we can examine the importance of cultural and ethnic heritage. Whether in war-torn areas or City streets, conflict resolution is an invaluable tool for finding ways to live together and collaborative greening endeavors are a viable mechanism for bringing people together – a way to find common ground within diverse communities. There is much to learn from those who usually do not share our table. For example, many refugees bring with them a cultural land ethic and many indigenous communities still retain traditional knowledge tied to the ways in which they have co-existed with the land. All that knowledge, largely untapped, offers a vast array of potential strategies that could help mitigate the degradation of our ecosystems, be they urban or otherwise. Our differences can serve as enrichment – of us as individuals, us as communities, as well as the environments that we depend on for our sustenance – in fact, for our very lives. If I think of how much trees have taught me about people, imagine what we could learn from each other, regardless of whether you are a red maple, growing in a valley in Tennessee and I am a blue pine, growing on a mountain in Afghanistan.

What is also clear to me is just how Urban and Community Forestry has taken diversity and used it as a way to find the common ground that facilitates sharing – of experience, of knowledge and of a willingness to steward that extraordinary resource – the tree. Quintessential U&CF is an acronym for the science and art of connecting the dots – trees, the landscapes they populate and the life they support including all of us.

On a note of hope. As I reflect on all the offerings in the previous chapter and assemble my closing thoughts, I do so against a backdrop of global disturbance, discord and angst.

In addition to the vast array of wars – against our own species which never bears the fruit promised but only the assurance of more wars, against the Earth, whose *fruits* we depend on for life itself – the human inhabitants of Planet Earth are experiencing a pandemic. Many of us are living in a state of imposed isolation. While this is quite contrary to us as social beings, it might offer a *teachable moment* – the time to reflect, not only on humanity's interdependence but our dependence on other species with which we co-exist and to which we are connected – seen and unseen, known and unknown, hiding in plain sight.

Touch the earth and listen to the rocks
for they remember
They know and remember
all that has come to pass here.
Lee Henderson

Has our urbanites' pandemic experience broadened our compassion for that many-species community in which we live? In retrospect, I can equate my own personal experience of imposed isolation with my professional grasp of what urban dwellers have imposed on another highly social organism – the tree and the landscapes we have asked it to populate.

Can we collectively use our own imposed isolation experience – what we each have endured during this pandemic time – to enhance our consideration of the tree and its connections? How the knowledge that you can take the tree out of the forest but you can never take the forest out of the tree must inform our understanding of how trees' genetic capacity dictates their communal needs. How accommodation of those communal needs can enable the what that our trees were meant to be, the what that we are so desperate for them to be and the what that all of us can do to ensure that happens.

In thinking of all the benefits that could be realized, I would like to take Joseph Beuys' most poignant thought "You don't make war when you are planting trees…" and expand on it so that we are not just planting trees, we are growing them and by growing trees, we are growing our Urban and Community Forest ecosystem for collective well-being, growing our connections to the Earth that sustains us so that we can also be sustained, growing our embrace of the importance of diversity in all things, enhanced through peaceful coexistence with Mother Earth and all her children. Or, from the Haudenosaunee governance perspective, that peace and life are synonymous and, in order to have life, one must sustain and be sustained, individually and collectively.

They are beautiful in their peace, they are wise in their silence.
They will stand after we are dust.
They teach us, and we tend them.
– Galeain ip Altiem MacDunelmor

Fig. 6.1 SINDROME DEL ESPEJO Montevideo IV. 2012. Pablo S Herrero y David de la Mano

References

Asia Brief, East Asia Division (2013) Community forestry in Bhutan – putting the people Back in charge. Schweizerische Eidgenossenschaft

Fisher DR, Svendsen ES, Connolly J (2015) Urban environmental stewardship and civic engagement: how planting trees strengthens the roots of democracy. Routledge, London (4)

Kardan O et al (2015) Neighborhood greenspace and health in a large urban center. Sci Rep 5:11610. https://doi.org/10.1038/srep11610

Koeser AK et al (2016) The cost of not maintaining trees: findings and recommendations from an international symposium and Summit Arboricult Urban For 42 (6):377–288 (4)

Mulcahy K, Zürcher N (2007) Tree shopping guide – selecting quality in the nursery.pdf. NYC Root Zone

Nowak DJ, Greenfield EJ (2012) Trees and impervious cover in the United States. Landsc Urban Plan 107:21–30

Nowak DJ, Noble MH, Sisinni SM, Dwyer JF (2001) People & trees: assessing the US Urban Forest resource. J For 99:37–42

Shigo AL (1991) Modern arboriculture: a systems approach to the care of tree and their associates. Shigo and Trees Associates, Durham, NH

Shigo AL (1996) How trees survive. TCI, VII, 2, (1)

Sitzia T, Campagnaro T, Weir RG (2016) Novel woodland patches in a small historical Mediterranean city: Padova, Northern Italy. Urban Ecosyst 19:475. https://doi.org/10.1007/s11252-015-0475-3

Sommer R, Learey F, Summit J, Tirrell M (1994a) Social benefits of resident involvement in tree planting: comparison with developer-planted trees. J Arboric 20(6):323–328 (4)

Sommer R, Learey F, Summit J, Tirrell M (1994b) The social benefits of resident involvement in tree planting. J Arboric 20(3):170–175 (4)

Vogt J et al (2015) The costs of maintaining and not maintaining the Urban Forest: a review of the urban forestry and arboriculture literature. Arboricult Urban For 41(6):293–322 (4)

Watkins SL (2017) Does collaborative tree planting between nonprofits and neighborhood groups improve neighborhood community capacity? Cities 74:83–99. https://doi.org/10.1016/j.cities.2017.11.006 (4)

Watson G (2001) A study of CTLA formula values. J Arboric 27(6):289–297

Watson GW, Himelick EB (1997, 2013) Principles and practice of planting trees and shrubs. International Society of Arboriculture

N. Zürcher, *Connecting Trees with People*, Future City 16,
https://doi.org/10.1007/978-3-030-94534-3

Westphal LM (2003) Urban greening and social benefits: a study of empowerment outcomes. J Arboric 29(3):137–147 (4)

Zürcher N (2007) Noteworthy requirements for conifer and hardwood selection.pdf. NYC Root Zone

Zürcher N (2017) Assessing the ecosystem services deliverable: the critical role of the urban tree inventory. In: Pearlmutter D et al (eds) The Urban Forest: Cultivating Green Infrastructure for People and the Environment. Future city, vol 7. Springer, Cham. https://doi.org/10.1007/978-3-319-50280-9_11

Zürcher N (2021) In consideration of the tree: the importance of structure and function in the realization of ecological Design. In: Catalano C, Andreucci MB, Guarino R, Bretzel F, Leone M, Pasta S (eds) Urban Services to Ecosystems. Future City, vol 17. Springer, Cham. https://doi.org/10.1007/978-3-030-75929-2_27

Index

N. Zürcher, *Connecting Trees with People*, Future City 16,
https://doi.org/10.1007/978-3-030-94534-3